# Scientific Seed Production of Horticultural Crops

# Scientific Seed Production of Horticultural Crops

**M. Jayanthi**
**S. Sumathi**
**B. Venudevan**
Tamil Nadu Agricultural University
Coimbatore, Tamil Nadu

NEW INDIA PUBLISHING AGENCY
New Delhi – 110 034

**NEW INDIA PUBLISHING AGENCY**
101, Vikas Surya Plaza, CU Block, LSC Market
Pitam Pura, New Delhi 110 034, India
Phone: + 91 (11) 27 34 17 17 Fax: + 91 (11) 27 34 16 16
Email: info@nipabooks.com
Web: www.nipabooks.com

Feedback at feedbacks@nipabooks.com



ISBN 978-93-87973-35-0



Composed and Designed by NIPA

# Preface

Horticultural crops occupy an important place in diversification of agriculture and have played a pivotal role in nutritional security. With the changing paradigms of food and nutritional securities, the consumption of vegetables have attained tremendous importance. The book has been prepared keeping in view the course requirements of seed science and technology for undergraduate students at various institutions. The large volumes of materials are available for this subject. An attempt has been made to consolidate the scattered information and presented in a simple format. This publication describes the procedures involved in the production of quality seeds of 40 horticultural crops. The techniques involved in the certified and foundation stages of seed production from seed selection to storage including the quality standards and appropriate storage methods are explained in detail in this publication. From this the readers will get knowledge about seed production of horticultural crops. It is hoped that the publication will be useful for vegetable seed growers, research workers, teachers, students, planners, NGOs and extension personnel. We hope that this book "**Scientific Seed Production of Horticultural Crops"** will be helpful to our readers of this subject.

**Authors**

# Contents

## Section 2: Ornamentals



## Section 3: Medicinal Plants

## Section 4: Seed Spices



## Section 5: Plantation Crops



## Section 6: Minor Fruits

# Section 1: Vegetables

# 1

# Tomato (*Lycopersicum* Linnacus Karst. Solanaceae)

Tomato is one of the most popular and widely grown nutritious vegetable in the world. Tomato (*Lycopersicum esculentum* Linnacus Karst.) belonging to the family Solanaceae is one of the most nutritious and remunerative vegetable crop. It is the world's important vegetable next to potato. The crop is cultivated throughout the year. In Northern plains, the crop is cultivated during autumn, spring and summer. In South India, the crop is grown in June – July, October – November and January – February. The optimum season for seed production in southern India is October – December. Nursery raised in late October and transplanted in the first week of December will produce a good seed crop.

## Floral Biology and Pollination Habit

Tomato inflorescence or flower cluster is borne laterally in small forked raceme cyme. The number of flowers per cluster in most cultivars varies from 4 to 5 and some times more. In most commonly grown field varieties about 2 to 4 flowers set fruits within each cluster.

Tomato flower has a 5 to 10 parted calyx which persists until fruit matures. The yellow petals are united in a short tube with five or more lobes which are often recurved. The five stamens are attached to the base of the corolla tube. The long anthers are partly united in the form of a cone surrounding the pistil. The latter consists of a multicelled ovary and a long slender style reaching the tip or projecting from the staminal cone as much as 2 mm with a capitate, single narrow or somewhat bulbous stigma. The buds, flowers, and fruits develop progressively within an individual cluster. There is no definite flowering peak in tomatoes. Anthesis appears to be correlated with temperature and soil moisture.

Tomato is normally self-pollinated crop. Self-fertilization being favoured by the position of the receptive stigma within the cone of anthers and the normal pendant position of the flower. Though the stigma is receptive at the time of anthesis anthers do not dehisce until about 24-48 hours later. Cross-pollination of tomato flowers to the extent of about 5 per cent may occur through insects.

### Method of seed production

The crop should be raised in isolation from the fields of other varieties. The isolation distance maintained between the fields of other varieties and the fields of the same variety not conforming to the varietal purity requirements for certification is 50 metres for foundation and 25 metres for certified seed production.

### Land selection

The land selected should not be cultivated with tomato in the previous season and free from volunteer plants. The soil should be fertile, free from soil borne diseases and with good drainage facility.

### Seed selection and treatment

Certified seeds should be obtained from an authorised source. Seeds should be healthy, free from disease and pest infection. Remove the broken, coloured seeds and use uniformly graded seeds. Seed rate is 200 gm/acre (500 gm/ha). Selected seeds should be treated appropriately to prevent the crop from seed borne diseases. Seeds should be soaked in a fermented mixture of buttermilk (3 days old) and water in a 1:4 ratio for six hours and shade dried before sowing. The practice is applicable only for the seeds which are 6 to 12 months old. The seeds should be treated with *Trichoderma viride* and *Pseudomonas fluorescens* (@ 5 g/100g of seeds). This will help in the control of early blight and other pathogens.

### Varieties

**Indeterminate varieties:** Pusa Ruby, Solan Gola, Yaswant (A 2), Sioux, Marglobe, Naveen, Ptom 9301, Shalimar 1, Shalimar 2. Angurlata, Solan Bajr, Solan Sagun, Arka Vikas. Arita Saurbh.

**Determinate varieties:** Roma (EC 13513), Rupali, MTH 15, Ptom 18, VL 1, VL 2, HS 101, HS 102, HS 110, Pusa Early Dwarf, Pusa Sheetal, Floradade, Arka Meghli, Co.1, Co.2, Co.3 (Marutham), PKM.1, Py1,

**Hybrids:** COTH 1, Pant Hybrid 2, Pant Hybrid 10, Kt 4. Pusa Hybrid 1 4, Arka Shreshta, Arka Vardan, Arka Abhijit, Navell 1 &2 (Sandoz), Rupali, Sonali, MTH6

### Nursery preparation and sowing

Seeds are sown in the nursery and then the seedlings are transplanted in the main field. Nursery beds of 2 – 2.5 metre long and 1 – 1.25 metres wide @ 10 numbers per acre (25 numbers per hectare) should be laid to raise the required seedlings. The beds should be raised 15 – 20 cm from the ground level. The bed should be covered with a layer of farmyard manure and sand in equal proportion. Addition of farmyard manure should be @ 4 kg/m2. Neem cake and groundnut cake (@ 2 kg/cent) can also be added to enrich the nursery soil. Dusting of wood ash on the seedlings in the nursery acts as an insect repellent and protects the young plants from the pest and disease attacks. It also serves as a good source of mineral nutrients. Treated seeds should be sown in the nursery beds in rows with 3 – 4 cm spacing. Soon after sowing irrigate the beds using a rose can and cover the beds using paddy straw or coconut fronds.

## Transplanting

The seedlings are transplanted to the main field 4 –5 weeks after sowing, preferably in the evening. At the time of transplanting the plant should be about 7.5 – 10 cm in height and with a sturdy stem. The roots of the seedlings should be soaked in asafoetida solution (100 gms in 5 litres of water) for 15 – 30 minutes before transplanting. This prevents the soil borne bacterial diseases. After uprooting, the roots of seedlings can also be dipped in cow dung and cow's urine slurry / Cow Pat Pit / Amrit Pani / Panchagavya overnight before transplanting in the field. This helps in better root growth and early establishment. The treated seedlings should be transplanted following 75 x 60 (or) 45 x 30 cm spacing. A well prepared seed bed with 4–5 ploughing is necessary for transplanting tomato. The seedlings are transplanted on the sides of the ridges.

## Nutrient management

Farmyard manure is applied @ 10 tonnes/ acre (25 tonnes/ha) during first ploughing and incorporated into the soil. Green manure with crops like sunhemp (*Crotalaria juncea* L.), cowpea (*Vigna unguiculata*), Daincha (*Sesbania bispinosa* Jacqs.) and cluster bean (*Cyamopsis tetragonoloba* (L.) Taub. ) can also be used to substitute the farmyard manure. Neem cake should be applied @ 60 – 100 kg/acre (150-250 kg/ha) as a basal dose to prevent nematode attack. Top dressing should be given with groundnut cake @ 30 – 40 kg/acre (80-100 kg/ ha) after 40 days of sowing. This will help in increasing the yield as well as the size of the fruits.

## Weed management

Weeding during the initial stages of plant growth is very necessary. Manual weeding is most preferred. Weeding at 45 days after transplanting is very critical. The plants require frequent shallow hoeing especially during the first four weeks after transplanting. This facilitates soil aeration for proper root development. Hoeing can be done to loosen the soil after every irrigation. Earthing up should be done twice. Irrigation First irrigation is done immediately after sowing.

Subsequent irrigation should be done once a week or 10 days depending on the soil moisture. Irrigation during flowering and fruit setting stages are very crucial.

## Major insect-pest and diseases of tomato

| Insects | Control Measures |
| --- | --- |
| Tomato fruit warm | Malathion 50 EC or Endosulfan 35 EC @ 2 ml/litre of water should be applied as foliar spray. |
| Epilachna beetle | Malathion/Fyfanon/Zithiol 50 EC @ 2 ml/litre of water can be sprayed. |
| Jassids | Monocrotophas 50 EC @ 2 ml/litre or Cypermethrin 10 EC @ 1 ml/litre of water to be sprayed and repeated at fortnight interval. |
| Aphids | Malathion 50 EC or Dimacron 50 EC 2 ml/litre of water as foliar spray. |
| Mealy bug | Fenthion imidacloprid 50 EC @ 2 ml/litre or Malathion/ Phosphamidon 100 EC @ 1 ml/litre of water to be sprayed. |

| Diseases | Control Measures |
| --- | --- |
| Damping off (*Phythium/Phytophthora* sp) | Soil sterilization by formaldehyde with 50 times water upto 4 inch soil depth. Seed beds should be kept clean and well drained. Seeds should be treated with captan 81 g per kg seeds. Spraying of 1% Bordeaux mixture at an interval of 8- 10 days of sowing. |
| Bacterial wilt *(Ralstonia solanacearum)* | Use of resistant varieties.The land should be kept clean and well drained. High degree of control of bacterial wilt in Indonesia by spraying 20 ppm Streptomycin sulfate (Oxytetracycline) at 4 and 7 days intervals was reported. |
| Tobacco mosaic virus( TMV) | Roguing out of mosaic affected plants as soon as observed. Control of the vector (white fly) by spraying Phosphamidon or Imidacloprid @ 1 ml/litre of water. |
| Lateblight *(Phytophthora infestans)* | Use of resistant varieties Spraying of fungicides, Dithane, 2-78 and Curzate @ 2 g/litre of water is effective. |
| Early blight*(Alternaria solani)* | Use of resistant varieties Spraying of Carbendazim/ Bairstin @ 2.5/Kg seed is reported to be effective. Spraying of Cystene @ 150 mg/litre of water was found effective. |

## Roguing

Roguing should be done from early vegetative phase upto three fruit stage. The plants that are morphologically different from other plants should be rogued off during vegetative stage. During fruiting stage based on the colour and shape of the fruit the off-types are rogued off. Maximum percentage of off-types permitted at the final inspection is 0.10% for foundation seed production and 0.20% for certified seed production.

## Field inspection

A minimum of three field inspections should be done from flowering to harvesting stage by the Seed Certification Officer. The first inspection is done before flowering, second during flowering and fruiting stage and the third during mature fruit stage or prior to harvest.

## Field standards

| | Foundation seed | Certified seed |
|---|---|---|
| Isolation distance | 50 m | 25 m |
| Off-types | 0.10% | 0.20% |
| Seed borne diseases affected seeds | 0.10% | 0.50% |

## Planting ratio

For hybrid seed production, the female and male parents are normally planted in the ratio of 12:1 or 12:2.

## Crossing (for hybrids)

In tomato the hybrid seed production is normally done by 'Emasculation and Hand Pollination'. However use of chemical hybridizing agents (MH 1000 ppm) or CMS lines are also practiced.

## Hybrid seed production

Emasculation is done before the anthers are mature and the stigma has become receptive to minimize accidental self-pollination. Thus emasculation is generally done in the evening, between 4 PM and 6 PM one day before the anthers are expected to dehisce or mature and the stigma is likely to become fully receptive.

Emasculate the bud by hand with the help of needle and forceps. Remove the calyx, corolla and staminal column or anthers, leaving gynoecium *i.e.*, stigma and style intact in the flower. Emasculated flowers should be covered immediately with red coloured paper cover to protect against contamination from foreign pollen and also for easy identification of emasculated bud during dusting. Remove the red paper cover of the emasculated bud and dust the pollen gently over the stigmatic surface using cotton or camel brush, *etc.* After dusting, the emasculated flowers are again covered with white or other coloured paper cover for two to three days. Pollen collected from one male flower can be used for dusting 5 to 7 emasculated flowers.

## Harvesting

Harvesting is done once the fruits are physiologically mature and turns from green colour to orange or red. The fruits of the lower three hands of each plant is the best for seed extraction. The fruits that should be harvested are those that are ripe just beyond the eating stage.

## Seed extraction and processing

The seeds with the pulp of the mature fruits should be squeezed into a jar and left in a warm spot for two to three days and allowed to ferment. Then the whole mass should be poured through a sieve, and the seeds should be rubbed and washed.

Alternatively, lemon juice can be used for seed extraction in place of corrosive hydrochloric acid (which is commonly used). The seeds should be treated with the juice @ 20 lemons / kg of wet seeds for 2-3 hours.

Seeds can also be extracted from the ripe fruits by squeezing the fruits on well-spread rice bran (@ 1 kg rice bran for 1 kg seed). After thorough mixing and drying for 24-48 hours, the bran is separated from the mixture by hand winnower.

For large scale seed extraction we can use the tomato seed extractor developed by Tamil Nadu Agricultural University. The seeds extracted by this machine may again be treated with commercial Hydrochloric acid @ 2-3 ml/kg seed with equal volume of water for 3-5 minutes with constant stirring. And then seed should be washed with water for to four times.

## Drying and storage

The extracted seeds are dried in the shade for a day or two before storage to attain a moisture level of 8%. The dried seeds should be packed in a cloth or moisture proof containers and stored in a dry and cool place. Under optimum conditions the properly dried seeds can retain the viability for 2-3 years.

**Seed Yield** : 100-120 Kg/ha

## Seed standard

The percentage of minimum physical purity of foundation and certified seeds should be 98% with a minimum of 70% of germination capacity and 8% of moisture content. The presence of inert matter should not exceed 2.0%

# 2

# Brinjal (*Solanum melongena* Linnaeus Solanaceae)

The brinjal, eggplant or aubergine (French name) has originated in the Indian sub-continent and China (Thomson and Kelly, 1957, Purewall, 1957 and Martin and Rhodes, 1979). Brinjal is an important vegetable crop of the Far East, Bangladesh, India, China and the Philippines. Brinjal belongs to the family Solanaceae and is known under the botanical name *Solanum melongena* L. (2n = 24). There are 3 main botanical varieties under the species *melongena* (Chowdhury, 1976). The round or egg-shaped cultivars are grouped under var. *esculenturn*, the long slender types are included under var. *serpentintum* and the dwarf brinjal plants are put under var. *depressum*.

## Floral Biology and Pollination Habit

Brinjal flowers are large, violet coloured and solitary or in clusters of two or more. Flower consists of calyx : sepals 5, united, persistent; corolla : petals 5, united, usually cup shaped; Androecium : stamens 5, alternate with corolla; Gynoecium: carpels are united, ovary superior.In most varieties the perfect flowers are borne singly and opposite the leaves. The stamens dehisce at the same time the stigma is receptive so that self-pollination is the rule although there is some cross-pollination by insects. Depending on the length of styles, four types of flowers are reported in brinjal, 1) long-styled with large ovary, 2) medium-styled with medium size ovary, 3) Pseudoshort-styled with rudimentary ovary and 4) true short-styled with very rudimentary ovary. It is observed that long and medium-styled flowers produce fruits whereas pseudo-short and short-styled flowers fail to set fruits. The anthesis and dehiscence in brinjal are mainly influenced by the daylight, temperature and humidity. Usually anthesis starts from 7-30 A. M. and continues upto 11. A. M. Peak time for anthesis is 8-30 to 10-30 A.M. The pollen dehiscence starts from 9-30 to 10 A.M.

### Method of seed production

The isolation distance maintained between the fields of other varieties and the fields of the same variety not conforming to the varietal purity requirements for certification is 300 metres for foundation and 150 metres for certified seed production.

### Stages of seed production

Breeder seed → Foundation Seed → Foundation Seed II → Certified Seed.

### Land selection

The land selected should be free from volunteer plants and objectionable weeds. The land should be fertile, rich in organic matter with good drainage facility.

### Seed selection and treatment

Certified seeds should be obtained from an authorised source. Seeds should be healthy and free from disease and pest infection. Remove the broken, coloured seeds and use uniformly graded seeds. Seed rate is 160 gms/acre (400 gms/ha). Selected seeds should be treated appropriately to prevent the crop from seed borne diseases. Seeds should be soaked in a solution of cow's urine (1 part cow's urine + 5 parts of water) for 30 minutes prior to the sowing. This will inhibit the seed borne diseases like fruit rot and die back (or) Seeds should be bundled using a thin cotton cloth and soaked in the bio gas slurry for 12 hours prior to the sowing. This will kill all the disease causing microbes and also enhance the seed vigour. Treat the seeds with *Trichoderma viride* @ 4 gms/kg of seeds.

### Varieties

**Long type** – Kashi Taru, Pusa Purple Long, Pusa Purple Cluster, Pant Samrat, Azad Kranti, Pusa Anupam, Punjab Sadabahar, Punjab Barsati, Arka Nidhi, Arka Sheel, Arka Keshav, Utkal Keshari, Swetha.

**Round type** - Pusa Purple Round, Pusa Kranti, Pusa Uttam, Pusa Upkar, Pant Rituraj, Kalyanpur T-3, Azad B-1, Azad B-2, Hisar Shyamal, Jamuni Gola, Punjab Bahar, Aruna, MDU-1, Swarna Mani, Utkal Tarini

**Green brinjal** - Green Brinjal Long-1, CO-1, Phule Harit, Arka Kusumakar, Arka Shirish, Utkal Madhuri

**Hybrids:** CoBH1,Arka Navneet (IIHR 22 1 x supreme), Pusa H 5, Pusa H 6, MHB 10, MHB 39 (Mahyco), Azad Hybrid.

**Season :** The brinjal seed production can be taken up in the following 2 seasons. May June and December January.

## Nursery preparation and sowing

Seeds are sown in the nursery and the seedlings are then transplanted to the main field. Nursery beds of 2 – 2.5 metre long and 1 – 1.25 metre wide @ 10 numbers per acre (25 numbers per hectare) should be laid to raise the required seedlings. The beds should be raised 15 – 20 cm from the ground level.

The bed should be covered with a layer of farmyard manure and sand in equal proportion. Addition of farmyard manure should be @ 4 kg/m$^2$. Neem cake and groundnut cake (@ 2 kg/cent) can also be added to enrich the nursery soil. Dusting of wood ash on the seedlings in the nursery acts as an insect repellent and protects the young plants from the pest and disease attacks. It also serves as a good source of mineral nutrients.

Treated seeds should be broadcasted or sown in the nursery beds in rows at 2 cm depth with 3 – 4 cm spacing. Soon after sowing irrigate the beds using a rose can and cover the beds using paddy straw or coconut fronds

## Transplanting

The seedlings are transplanted to the main field 4 – 5 weeks after sowing. At the time of transplanting the seedling should be about 12 – 15 cm tall. The roots of the seedlings should be soaked in asafoetida solution (100 gms in 5 litres of water) for 15 – 30 minutes before transplanting to prevent the soil borne bacterial diseases. After uprooting, the roots of the seedlings can also be dipped in cow dung and cow's urine slurry/Panchagavya overnight before transplanting in the field. This helps in better root growth and early establishment.

The treated seedlings should be transplanted to the main field. The spacing followed is 60 x 60 cm for non-spreading types and 90 x 60 for spreading type. The main field should be ploughed thoroughly for 4 - 5 times to get proper tilth before transplanting. The seedlings are transplanted on the sides of the ridges.

## Nutrient management

Farmyard Manure or compost is applied @ 10 tonnes/acre (25 tonnes/ha) before last ploughing and incorporated into the soil. Neem cake should be applied @ 60 – 100 kg/acre (150 - 250 kg/ha) as a basal dose to prevent nematode attack. Top dressing should be given with groundnut cake @ 30 – 40 kg/acre (80 - 100 kg/ha) after 40 days of sowing. This will help in increasing the yield as well as the size of the fruits.

## Weed management

Weeding is most important during the early stages of the crop. About 3 – 4 hoeing and weeding are required for effective control of weeds. Manual weeding is most preferred. The crop should be hoed once in 15 days from 30th day after transplanting for two times to get proper aeration in the soil. Weeding should be done on 20th and 45th day after transplanting.

## Irrigation

First irrigation is done immediately after sowing. Subsequent irrigation should be done once a week or 10 days depending on the soil moisture. Irrigation during the flowering and fruit setting stages are very crucial

## Major insect-pest and diseases of brinjal

| Insects | Control Measures |
|---|---|
| Fruit and shoot borer | Pirimiphos-methyl 50 EC @ 2 ml/liter of water or Cypermethrin 10 EC @ 1 ml/liter of water should be applied as foliar spray. Spraying should be repeated at an interval of 7-14 days where necessary. |
| Epilachna beetle | Monocrotophos /Fyfanon/Zithiol 50 EC @ 2 ml/litre of water can be sprayed. |
| Jassids | Nuvacron 50 EC @ 2 ml/litre or Cypermethrin 10 EC @ 1 ml/ litre of water to be sprayed and repeated at fortnight interval. |
| Aphids | Malathion 50 EC or Dimacron 50 EC 2 ml/litre of water as foliar spray. |
| Mealy bug | Imidacloprid 50 EC @ 2 ml/litre or Malathion/ Phosphamidon100 EC @ 1 ml/litre of water to be sprayed. |

| Diseases | Control Measures |
|---|---|
| Little leaf | Roguing out of affected plants, control of vectors by foliar spray of Methyl Parathion/Oxydemeton-methyl @ 1 ml/liter of water. |
| Damping off | Soil sterilization by Formaldehyde with 50 times water upto 4" soil depth. Seed treatment with Captan @ 1 g per kg seed. Spray captan @ 2 ml/liter of water. |
| Bacterial wilt | Proper crop rotation can reduce the infestation. Use of resistant varieties.Grafting on resistant solanum rootstock. |
| Phomopsis blight | Seed treatment with Ceresan/ Agrosan Spraying Bordeaux mixture (4:4:50) |
| Root knot nematode | Use of resistant varieties Treatment of soil with Chloropicrin before 15 days of sowing. |

## Roguing

Roguing should be done from early vegetative stage to flowering and fruiting stage. The offtypes are identified based on the morphological characteristics like plant type, shape and colour of the leaves and flowers, presence of thorns etc. The plants that are different from the other plants and diseased plants should be rogued off periodically. The maximum percentage of offtypes permitted at the final inspection is 0.10% for foundation seed production and 0.20% for certified seed production.

### Field inspection

A minimum of three field inspections should be done from vegetative to fruiting stage by the Seed Certification Officer. The first inspection is done at vegetative stage to check the isolation distance, presence of volunteer plants and diseased pants based on the physical appearance and other requirements. The second and third inspections should be done during flowering and fruiting stage and the off-types are identified based on the flower colour, fruit shape etc., and removed. The third inspection at maturity stage will also estimate the yield.

### Field standards

| | Foundation seed | Certified seed |
|---|---|---|
| Isolation distance | 300 m | 150 m |
| Off-types | 0.10% | 0.20% |
| Seed borne diseases affected seeds | 0.10% | 0.20% |

### Hybrid seed production

The planting ratio of female and male parents adopted for hybrid seed production is normally 5:1 or 6:1.For production of hybrid seeds, crossing programme is done using emasculation and dusting methods as followed in tomato.

### Harvesting

Harvesting is done once the fruits are physiologically mature. In brinjal, fruits are allowed to mature beyond the edible stage for seed purpose. The physiological maturity of the fruits is identified by change in colour. The mature fruits of different varieties will vary in colour from yellow to dull purple. The matured fruits are harvested by hand picking and hung in sheds until their colour dulls.

### Seed extraction and processing

The selected matured fruits are cut into pieces and crushed to extract the seeds. Then the pulp around the seeds are separated by washing and sieving. In other way, fruits should be cut into cubes and put in to a blender on slow speed with water. The masses and the pulp float on the surface can be removed and the seeds which settle at the bottom should be collected, washed and dried. In general, the extraction process should be done in the morning hours to make sure that the seeds are at least half dried by evening in order to avoid the danger of germination.

## Drying and storage

The extracted seeds should be thoroughly washed and dried on a sieve in the shade for a day or so before storage to attain a moisture level of 8%. For longer storage period the seeds should be dried to a moisture level of 6%. The dried seeds should be packed in paper bags and hung for a couple of weeks before storage. If seeds are packed in moisture proof polythene bags (700 gauge polythene bags) and stored in cool dry place it can be stored for a long time.

**Seed Yield** : 100-200 Kg/ha

## Seed standards

The percentage of minimum physical purity of the foundation and certified seeds should be 98% with a minimum of 70% of germination capacity and 8% of moisture content. The presence of inert matter should not exceed 2.0%.

# 3

# Chilli (*Capsicum* Sps. Solanaceae)

Chilli or Pepper (*Capsicum* Sps.) originated in South America and spread into the New World tropics before subsequent introduction to Asia and Africa. Chillies are now widely grown throughout the tropics, sub-tropics and warmer temperature region of the world. The genus *Capsicum* is a member of the family Solanaceae.

## Floral Biology and Pollination Habit

Chilli flower is normally solitary but occasionally borne in small cymes of leaf axils. The calyx is five lobed and corolla is five-parted and white, but occasionally purple in colour. The five stamens attached to the base of corolla are separated. The bluish anthers dehisce by splitting longitudinally. The single style is usually longer than the stamens and stigma is club-shaped. The ovary generally has three locules. Pepper tend to blossom and set fruit earlier under short day conditions.

Fruits of the capsicum is a pod like berry with a short, thick peduncle. The shape of the fruit varies from a flattened oblate to long slender and tapering, the size also ranges from very small to large fruit of sweet pepper. Seeds within the fruits mature as the fruit ripens. The seed is borne in a compact formation on the plancentae and usually at the basal end of the fruit. Peppers are generally self-pollinated, but some cross-pollination can occur between and within the cultivars of the two species (*C. annuum* and *C. frutescens* Linnaeus).

## Method of seed production

Capsicum and chilli are often self-pollinated crops, but cross-pollination occurs to the extent of 7 - 36% mainly through insects. Seeds should be allowed to set by self-pollination. The isolation distance maintained between the fields of other varieties and the fields of the same variety not conforming to the varietal purity requirements for certification is 500 metres for foundation and 250 metres for certified seed production.

### Seed production stages

Breeder seed → Foundation seed → Certified seed

### Land selection

The land selected should be free from volunteer plants and objectionable weeds. There should be at least two years interval between the related crops cultivated in the selected land. The soil should be fertile, free from soil borne diseases and with good drainage facility.

### Seed selection and treatment

Certified seeds should be obtained from an authorised source. Seeds should be healthy and free from disease and pest infection. Remove the broken, coloured seeds and use uniformly graded seeds. Seed rate is 400 gms/acre (1 kg/ha). Selected seeds should be treated appropriately to prevent the crop from seed borne diseases. Seeds should be soaked in a solution of cow's urine (1 part cow's urine + 5 parts of water) for 30 minutes prior to the sowing. This will inhibit the seed borne diseases like fruit rot and die back. (or) Seeds should be bundled using a thin cotton cloth and soaked in the bio gas slurry for 12 hours prior to the sowing. This will kill all the disease causing microbes and also enhance the seed vigour. Treat the seeds with *Trichoderma viride* @ 4 gms / kg of seeds (or) Treat the seeds with biofertilizers @ 1 kg/acre of seeds. Mix the biofertlizers with rice gruel and then mix it with seeds. Dry the seeds under shade for 30 minutes before sowing.

### Nursery preparation and sowing

Seeds are sown in the nursery and the seedlings are then transplanted to the main field. Nursery beds of 2 – 2.5 metre long and 1 – 1.25 metre wide @ 10 numbers per acre (25 numbers per hectare) should be laid to raise the required seedlings. The beds should be raised 15 – 20 cm from the ground level. The bed should be covered with a layer of farmyard manure and sand in equal proportion. Addition of farmyard manure should be @ 4 kg/m$^2$. Neem cake and groundnut cake (@ 2 kg/cent) can also be added to enrich the nursery soil. Dusting of wood ash on the seedlings in the nursery acts as an insect repellent and protects the young plants from pest and disease

### Land Preparation and Planting

The land is ploughed 3-4 times to get a fine tilth. Farmyard manure is incorporated during the last ploughing. For irrigated crop, ridges and furrows are made. Seedlings are transplanted 4-5 weeks after sowing at a spacing of 30 x 30 cm. Levy *et al.* (1983) reported that increased plant diversity resulted in less lateral branching making the fruits easier to harvest.

## Manures and Fertilizers

Apply 50 tones of FYM/ha for irrigated crop. Basal 0:70:70 kg of NPK and 50 kg of N at 15 days after transplanting and 50 kg N at 45th days after transplanting.

## Irrigation and Weeding

The uniform soil moisture is essential to blossom and prevent fruit drop. Generally 8-9 irrigations are given, depending upon rainfall, soil type, humidity and prevailing temperature. Two to three weeding and mulching are necessary to keep the field clear of weeds.

## Isolation

Chilli is considered as self-pollinated crop, but significant cross-pollination does occur if plants are placed together. A minimum distance of about 400m between two varieties is recommended.

## Major insect-pest and diseases of brinjal

| Insects | Control Measures |
|---|---|
| Aphids | Malathion 50 EC or Phosphamidon 50 EC 2 ml/litre of water as foliar spray. |
| Thrips | Malathion 50 EC@ 1 ml/litre of water should be sprayed |

| Diseases | Control Measures |
|---|---|
| Damping off | Soil sterilization by Formaldehyde with 50 times water upto 4" soil depth. Seed treatment with Captan @ 1 g per kg seed. |
| Anthracnose and Fruit rot (*Colletotrichum capsici*) | Seed treatment with Benlate or Bavastin @ 2 g/kg seed. Foliar spray of carbendazim followed by benomyle |
| Yellow mosaic virus | Use of resistant varieties. Roguing of the diseased plants |

## Roguing

Plants should be rogued based on the plant and fruit characters as a whole rather than the individual character. Off-types should be removed as soon as they are observed. When the fruits begin to show their final colour of red or yellow, occasional plants with off-colour fruits have to be removed. In addition to off-types, diseased plants are also to be removed.

### Field standards

| | Foundation seed | Certified seed |
|---|---|---|
| Isolation distance | 500 m | 250 m |
| Off-types | 0.10% | 0.20% |
| Seed borne diseases affected seeds | 0.10% | 0.50% |

### Harvesting

Harvesting is done once the fruits are physiologically mature and turns from green colour to red. The matured fruits are harvested by hand picking.

### Seed extraction and processing

The seeds are extracted from freshly harvested pods or from the dried pods after proper drying. The dried pods are taken in a gunny bag and beaten with a pliable bamboo stick to separate the seeds. The seeds are then cleaned by winnowing. The seeds from the fresh pods should be scraped out and dried in the shade for a few days. Seeds from fresh pods can also be extracted by following methods,

For large quantities of seeds, the ripe pods along with water should be blended in a blender at slow speed. Pulp will rise to the top and seeds will settle in the bottom. Seeds should be collected after decanting the water.

The seeds from the matured fruits should be squeezed into a jar and left in a warm spot for two to three days and allowed to ferment. Then the whole mass should be poured through a sieve, and the seeds should be rubbed and washed. Alternatively, lemon juice can be used for seed extraction in place of corrosive hydrochloric acid (which is commonly used). The seeds should be treated with the juice @ 20 lemons / kg of wet seeds for 2 - 3 hours.

Seeds can also be extracted from the ripe fruits by squeezing the fruits on well-spread rice bran (@ 1 kg rice bran for 1 kg seed). After thorough mixing and drying for 24 - 48 hours, the bran is separated from the mixture by a hand winnower.

### Drying and storage

The extracted seeds should be dried in the shade for a few days before storage to attain a moisture level of 8%. The dried seeds should be packed in cloth bags or moisture proof containers and stored in a dry and cool place. When the seeds are kept under the cold, dark and dry storage conditions, they will remain viable for upto five years.

## Seed Yield

Average seed yield varies from 50 to 80 kg per hectare.

## Seed standards

The percentage of minimum physical purity of foundation and certified seeds should be 98% with a minimum of 60% of germination capacity and 8% of moisture content. The presence of inert matter should not exceed 2.0%. The presence of other crop seeds and weed seeds should not be more than 5/kg for foundation seeds and 10/kg for certified seeds.

# 4

# Bhendi (*Abelmochus esculentus* Linnaeus Malvaceae)

Bhendi is known by many local names in different parts of the World. For example it is called Lady's finger in England, Gumbo in USA, bhendi in India Okra (*Abelmochus esculentus* L. 2n= 1 30) belongs to the family Malvaceae. Origin of okra is West Africa. Okra plant is an erect, herbaceous annual, 1-2 meter tall and forms fairly heavy tap root.

## Floral Biology

A flower bud appears in the axil of each leaf above 6th to 8th leaf depending upon the cultivar. Flowers solitary, axillary with about 2 cm long peduncle; epicalyx up to 10, narrow hairy bracteoles which fall before the fruit reaches maturity; calyx split longitudinally as flower opens; petals 5, yellow with crimson spot on claw; 5-7 cm long; staminal column united to the base of petals with numerous stamens; ovary superior, stigma 5-7m m, deep red. A flower bud takes about 22-26 days from initiation to full bloom. The time of anthesis varies with the cultivar, temperature and humidity and it ranges from 8 to 10 A. M. The dehiscence of anther occurs 15 to 20 minutes after anthesis. The dehiscence is complete in 5-10 hours for fertilization after pollination. The flowers remain open for a short time and they wither late in the afternoon. The stigma is receptive at the time of opening of flowers.

## Varieties

Co.1, Co.2, MDU.1, Parbhani Kranti, Arka Anamika, Pusa A 4, Pusa Savani

## Land Preparation, Manure and Fertilizer

Land should be thoroughly prepared by deep ploughing, harrowing, laddering etc.

### Manures and fertilizers

Apply 12.5 tons of FYM/ha before ploughing. Apply 150:75:75 kg NPK/ha, of which 50% of the N should be applied as top dressing in two split doses at flowering and 10 days later.

### Seed Sowing

Seed should be sown during (a) mid June-Mid July and (b) Mid February-mid March. Soil temperatures between 27-30°C help in quick and better seedling emergence. Seeds will not germinate below soil temperatures of 17°C. Seeds should be soaked in clean water for 24 hours before sowing. Seeds which will not absorb water during imbibition should be discarded. Seeds to be sown in lines and in small hills. Spacings of 60 cm. between lines and 30 cms between plants are to be maintained. 2/3 seeds should be sown per hill.

### After Care

- Only one healthy plant should be allowed per hill.
- Land should be kept clean by weeding and mulching.
- Irrigation is to be provided in dry season crop as and when required.
- In wet season crop, however, irrigation should only be given if there is no rain for a longer period.
- Earthing up along the lines to be provided at the time of top-dressing urea.
- Roguing is very important and should start as early as possible. All off types and plants affected by virus must be removed as soon as they are observed. The crop will normally require several roguings.

### Isolation

A minimum isolation of 500 m is desirable.

### Field inspection

A minimum of three field inspections should be done from flowering to harvesting stage by the Seed Certification Officer. The first inspection is done before flowering, second during peak flowering and fruiting stage followed by the third one during mature fruit stage or prior to harvest.

## Field standards

| | Foundation seed | Certified seed |
|---|---|---|
| Isolation distance | 500 m | 250 m |
| Off-types | 0.10% | 0.20% |
| Seed borne diseases affected seeds | 0.10% | 0.50% |

## Major pest and disease control measures

| Insects | Control Measures |
|---|---|
| Aphids | Malathion 50 EC or Phosphamidon 50 EC 2 ml/litre of water as foliar spray. |
| Leaf hopper | Malathion 50 EC@ 2 ml/litre of water should be sprayed |
| Shoot or fruit borer | Spraying of Bidrin or carbicron 100 EC @ 1 ml/litre of water |
| Shoot borer | Dicrotophos 8 or Phosphamidon 100 EC @ 1 ml/litre of water or Diazionon 50 EC@ 2 ml/litre |

| Diseases | Control Measures |
|---|---|
| Yellow mosaic virus | Protection of the crop from white fly, the vector by spraying Cypermethrin 50 EC@ 2 ml/litre of water. |
| Powdery mildew | Spraying of Bavistin @ 2 g/litre of water |
| Damping off | Seed treatment with 3 g Captan or Thiram per kg of seed before sowing. |

## Harvesting

Harvesting is done once the pods are physiologically mature. The physiological maturity of pods is identified by a change in colour from green to brown and by the drying of the pods. Pods should be harvested at the right time, since dried pods tend to dehisce (split open) with very little force.

## Threshing and processing

Harvested pods should be dried under the sun. Later, seeds should be removed from the peels of the pods by beating with stick. The separated seeds are then winnowed to remove the debris.

## Drying and storage

The seeds should be dried well before storage upto 10% of the moisture content. Under dry climatic conditions seeds can be stored for one year.

**Seed Yield :** 1.0-1.5 t/ha

## Seed standards

The percentage of minimum physical purity of foundation and certified seeds should be 99% with a minimum of 65% of germination capacity and 10% of moisture content. The presence of inert matter should not exceed 1.0% and the seeds of other crop varieties should not be more than 10/kg of foundation seeds and 20/kg of certified seeds.

# 5

# Cowpea (Vigna unguiculata L. Walp. Papilionaceae)

### Botanical description

Cowpea belong to genus Vigna and species *unguiculata* (L) Walp (Syn *V. sinensis* Savi) of the family Leguminosae sub family Fabacae.

### Floral biology

The flowers of cowpea are hermaphrodite. Self-pollination is effected though with some cross-pollination.

### Climate and Soil

Cowpea is a warm season crop, and therefore it can be grown both in spring and in rainy seasons in the plains of India. It cannot tolerate cold weather, heavy rainfall and water logging.

### Method and time of sowing

The seeds are dropped in the furrow in such a way that maintains distance approximately 10 to 15 cm in the rows which are at 40 to 60 cm apart for rainy season crop whereas summer crop is sown at the row distance of 25 to 30 cm.

### Seed rate

The requirement of seed for spring season crop is 20 to 25 kg/ha and for rainy season crop is 12-15 kg/ha.

### Nutrition

Cowpea responds well to an addition of manure and fertilizers. Application of 25 to 30 t/ha FYM improves the yield and quality of cowpea. About 20-25 kg nitrogen and whole dose of phosphorus (50-60 kg/ha) and potassium (50-60 kg/ha) are applied in soil during the last field preparation (6.1). Cowpea is highly sensitive to Zn deficiency. Application of 10 to 15 kg Zinc sulphate per hectare would be beneficial.

### Roguing

The seed crop of cowpea is rogued out for all off-types and diseased plants form the crop before flowering and during flowering. When the pods mature, at this stage off-types can be detected.

### Important Operations

In Cowpea the tendril are to be clipped off (pinching) for good seed setting. Spraying of NAA 40 PPM at flower initiation and at peak flowering stage will promote pod and seed setting.

### Pre-harvest Sanitation Spray

Two weeks before harvest endosulfan 0.07% should be sprayed twice at weekly interval to control pod borer and primary infestation of Bruchids.

The percentage of minimum physical purity of foundation and certified seeds should be 98% with a minimum of 75% of germination capacity and 9% of moisture content. The presence of inert matter should not exceed 2.0%.

The seed crop of cowpea is rogued out for all off-types and diseased plants form the crop before flowering and during flowering. When the pods mature, at this stage off-types can be detected.

### Field standards

| Factor | Foundation seed | Certified seed |
|---|---|---|
| Isolation distance | 50 m | 25 m |
| Off-types | 0.10 | 0.20 |
| **Plants affected by seed borne diseases | 0.10 | 0.20 |

Seed borne diseases are: Ashy stem blight, Anthracnose, Ascochyta blight, Cowpea mosaic

### Harvesting

The seed crop of cowpea matures in 75 to 125 days, depending upon the season and the variety. The pods turned into straw colour. Entire plant is harvested at the ground level and are allowed to dry in the field or heaped at one place in threshing floor for drying.

### Drying

The dried material is threshed by thresher or trampled. By winnowing all innert matter, chaff etc. are taken out. Cleaned seed of cowpea is spread on tarpaulin for drying till 10 percent moisture remained in seed.

### Seed yield

Seed crop of cowpea produces about 10-15 quintals of seed per hectare.

### Seed standards

The percentage of minimum physical purity of foundation and certified seeds should be 98% with a minimum of 75% of germination capacity and 9% of moisture content. The presence of inert matter should not exceed 2.0%.

# 6

# Lablab (*Dolichous lablab* Linnaeus Papilionaceae*)*

Lablab purpureus usually knowna s Dolichus bean, Hyacinth bean or Field bean belonging to the family fabaceae, one of the most ancient crops among cultivated plants. It is a bushy, semi-erect, perennial herb, showing no tendency to climb. It is mainly cultivated either as a pure crop or mixed with finger millet, groundnut, castor, corn, bajra or sorghum in Asia and Africa. It is a multipurpose crop grown for pulse, vegetable and forage. The crop is grown for its green pods, while dry seeds are used in various vegetable food preparations.

**Season:**Seed is very sensitive to weather. A grain crop can be sold for a lower price even if it damaged by rain. But a seed crop, if affected will lead to low seed quality. Hence selecting the right season is necessary. Generally, the seed should mature in cool dry climate. Seasons should be selected with this idea in mind. In Tamil Nadu the best seed crops are grown during March--April and September - October as the seeds mature during the cool months of January - February.

**Seed rate**

| Variety | Seed rate (kg /acre) |
|---|---|
| Co 3, Co 4, Co 6, Co 7 Co 8 | 1.5 |
| Co 9, Co 11, Co 12 | 8 |
| Co 10 | 10 |

**Seed treatment using fungicide**: Before sowing, the shrunken, shriveled, fungal infected and bruchid infected seeds must be removed and only good seeds must be sown. Seeds are treated with bavistin 2g/kg. The seeds are treated with Rhizobium 24 hours after fungicide seed treatment. Now-a-days organic fungicide like *Trichoderma viridi* is recommended for pulses at the rate of 4 g / kg of seed.

**Seed treatment using Rhizobium:** Two packets of Rhizobium will be needed to treat seeds of an acre. In order to make the Rhizobium stick to the seeds, we need a binder. The binder is prepared using rice gruel. This is prepared by adding 100 gm of rice to 500 ml of water and boiled until the rice becomes sticky. Three hundred milliliters of rice gruel is needed for preparing the rhizobial inoculation. The dried seeds are then used for sowing within 24 hrs of inoculation.

**Sowing :**Two seeds are sown at 2 cm depth for both bushy and climbing types.

**Selection of land:** A fertile and healthy seed plot will certainly produce quality seed. The field selected for seed production must not have been sown with motchai in the previous season. This is done to avoid volunteer plants that cause admixture. Fields continuously cultivated with lab lab may harbor wilt pathogen.Hence, such fields must be avoided wherever possible. Following the crop rotation will help to reduce endemic pathogen. Soil with neutral pH must be selected. Loam or clay loam soils are best "suited. Higher organic matter will lead the production of vigorous seed.

**Isolation :**The quality seed must be genetically and physically pure. Genetic purity can be maintained by preventing cross pollination with other undesirable varieties. This is achieved by isolating the seed crop. Isolation is the act of growing the seed crop away from a contaminant such that cross pollination is prevented. Lab lab is a self pollinated crop. The lab lab seed crop must be grown 25 m (75 feet) away from another vari-ety of lab lab.

**Preparation of land :**The land is made in to ridges and furrows of one and half feet wide with a plant spacing of half feet or beds and channels of 4' x 6' depending on cultivation practices.

**Irrigation:** Seed crop is very sensitive to irrigation. Seed fields must be constantly monitored for drought conditions. If not irrigated properly, pulse crop shed flowers. Water starved plants produce seeds that are hard and small with low vigor. Hence, pulse seed crop needs regular and sufficient irrigation. water is applied immediately after sowing followed by life irrigation on the third day. Then, irrigation is carried out whenever, the fields become dry. Irrigation during flowering, pod formation and seed development are must.

**Pruning techniques in climbing varieties** :Climbing varieties are allowed to climb over a wire netted trellis for better flowering and fruit formation. Two seeds are sown and only one healthy seedling is allowed to climb the trellis. As the plant reaches the top of the trellis, the terminal bud is nipped. This allows the prolific production of side shoots. As the side shoots advance, they are also nipped at 3' length so that the side shoots form into the trellis and not outside it. Branches arising from the main vine are removed periodically.

**Foliar feeding of fertilizers**: Some times, basal application of fertilizers alone is not sufficient for the seed crop. It is necessary that we provide nutrition during the heavy growth and seed formation period when there is a huge demand for nutrients. The nutrient needed by the fast developing pulse seeds are provided through foliar feeding. This provides rapid food source like sick human are provided glucose fluid through intravenous method. We have to prepare a DAP nutrient solution for spraying. This done by soaking 4 kg of DAP fertilizer in 20 litres of water overnight. The next day morning, the solution is filtered through a cloth. One litre of filterate is then added to the sprayer tank followed by adding 9 litres of water to fill the tank. The solution is then sprayed during the early morning or evening to avoid scorching. The crop must be irrigated immediately after spraying. This nutrient solution is sprayed 100 and 120 days after sowing for climbing, 45 days after sowing for bushy followed by another spray at 10 days later.

**Weed control** :Immediately after sowing and irrigation, Basalin herbicide is to be sprayed by dissolving 2 ml of Basalin per litre of water. The spraying of weedicide must be done within three days of sowing. If done later, it can harm the seed crop. Application of weedicide will control the early growing weeds, in order to control later emerging weeds, manual weeding done after 15 days will be useful. Some times, 600 ml of Basalin can be mixed with 20 kg sand (4 iron chatties of sand) and evenly spread on the field within three days of sowing. This reduces the cost of spraying.

## Pest and disease control

**Pests :**During the early stages, aphids may infect the crop and can be controlled by spraying Dimethoate or Phosphomidan 2 ml/litre of water. The major pest is *Heliothis*, which makes holes in young pods and eats the seeds. The worms can be collected and killed or sprayed with Oxydemeton-methyl Dimethoate or Phosphomidon at the rate of 2 ml per litre.

**Diseases :**During the growing phases of the seed crop, incidence of wilt can be seen. The affected plants turn brown and die. These plants can be rogued out as and when they appear. The affected area can be sprayed using 0.1 % Bavistin solution. Yellow mosaic can also be seen as the affected leaves turn yellow. Such plants are pulled out and destroyed followed by a spray of quinalphos 3ml/1 to ward of whitefly.

Powdery mildew is noticed by the presence of white powder deposits on the leaves. The Powdery mildew can be controlled by spraying Dithane M 45 @ 4g / litre.

Anthracnose disease which appears as round spots on leaves can be controlled by spraying Mancozeb 1% solution.

**Maintaining seed quality by rouging:** One of the most important aspects in seed production is thorough rouging. Rogue is defined as the presence of those plants that deviate from the char-acters described for the variety. Such rouges if left in the field tend to reduce the genetic purity of the seed crop and thus reduce market value of the re-sultant seed. Roguing is defined as the operation of removing rouges. In practice, all plants that do not obey the characteristics of the particular seed crop are to be removed along with diseased plants, other crop plants, weeds and insect affected plants during rouging operation

**Rouging during vegetative phase:** During vegetative phase *i.e.* during first 25 days, rouging is attempted based on plant characters like height of plant, leaf shape, size, venation, surface of leaf and presence or absence of hairs on plant surface. Plants showing wilt are also removed.

**Rouging during flowering phase:** The rouges are identified based on flower characteristics. During this phase when most of the plants are flowering, all those plants still in vegetative phase are also removed.

**Rouging during pod formation phase:** At this phase, the rouges are removed based on pod characteristics like length of pod, width, size, shape and colour are the major characteristics for roguing.

**Rouging during pod harvest phase:** Rouging is done prior to harvest based on seed characteristics like colour of seed, lustre and size of seed.

| Field standards | Foundation seed | Certified seed |
|---|---|---|
| Factor | Maximum permitted (%) | |
| Off types | 0.10 | 0.20 |
| Plants affected by seed borne disease | 0.10 | 0.20 |

**Harvest:** The mature pod comes to harvest generally 30 days from 50% flowering. Since, the flowering starts 120 and 45 days after sowing for climbing and bushy types, harvesting for seed is done when 70% of the pods turn straw coloured. Prior to harvest there is an important pest to be controlled. Bruchids are the major pests of stored pulse seeds.

**Pre-harvest sanitation spray:** Bruchids lay eggs on the surface, the grubs bore into the seeds and eat the cotyledon. The seeds thus store poorly and loose viability faster. It has been found that bruchids lay their eggs on the pods while in the field itself. Hence control of these pests must start from the field itself. The field carry-over of Bruchids can be controlled if the crop is sprayed using quinalphos 0.07% (2 ml of insecticide per litre) ten days prior to harvest. Upon ripening, the lab lab pods will turn from green to straw coloured. This is the right stage of harvest for seed purpose. Delaying will lead to infection by diseases, pests and sometimes seed vigor will be lost due to untimely rains. The first five harvest pods alone are used for seed extraction. During harvest, the shrunken, damaged and immature pods if any are removed.

**Processing :**One of the main characteristics of seed quality is the uniform size. Plumpy seeds are better than ill-filled puny seeds. Grading is one simple method by which we can separate the filled seeds from broken and puny seeds. Grading is done using round holed sieves. Such sieves are easily avail-able in the market. The sieve size for lab lab is 7mm. After sieving, broken, fungal infected, seed coat damaged seeds are removed.

**Seed treatment:** Seeds are generally affected by fungi during storage. To prevent this, seeds aretreated with fungicide be-fore storage. Lab lab seeds are treated using Bavistin at the rate of 4g/kg of seeds.

**Seed packing**: In case of short term storage, cloth bag can be preferred. Cloth bag is porous and hence can hold seeds with good vigour for short period only. However, cloth bags are cheaper and easily made using local tailor. Large quantity of seeds can be stored in gunny bag. If the seeds are to be stored for longer period, thick polythene bags can be used.

**Seed yield:** 12-15 Quintal/ha.

## Seed standards

The percentage of minimum physical purity of foundation and certified seeds should be 98% with a minimum of 75% of germination capacity and 9% of moisture content. The presence of inert matter should not exceed 2.0%.

# 7

# Cluster bean (*Cyamopsis tetragonoloba* Linnaeus Taub Papilionaceae)

**Botanical description :**Botanically, cluster bean is known as *Cyamopsis tetragonoloba* (L) Taub. (2n = 2x = 14) belongs to the family *Papilionaceae.*

**Floral biology :**Cluster bean is a self-pollinated crop having about 9 percent maximum natural out crossing. The plants are normally fully fertile, a few semi-sterile ones have also been reported.

**Isolation distance :**Cluster bean is a self-pollinated crop. Isolation distance of 50 metres for foundation and 25 metres for certified seed is essential.

**Sowing :**Cluster bean can be sown twice in a year, February - March in Northern plains and December - January in Southern plains.

**Seed rate :** In order to sow one hectare area about 30-40 kg seed of cluster bean is required.

**Nutrition :** The yield of cluster bean has been maximum when the crop was applied with 40 kg nitrogen and 60 kg phosphorus per hectare. The application of micronutrient to cluster bean crop proves very beneficial. Two sprays of molybdenum at 0.15 percent at 15 and 30 days after seedlings emergence give better yield.

**Roguing :**The first roguing should be done before flowering, the second one during flowering and fruiting stage and the third roguing at maturity.

**Harvesting :**When cluster bean pods attained full maturity (they turn grey in colour) the harvesting is done either by cutting entire plant at ground level or the plants are cut just below the first pod from ground level. Seed quality will be higher when seed crop is harvested after taking two pickings of fresh pods for vegetable use.

**Seed extraction :**It is done by threshing followed by cleaning (winnowing).

**Drying :**Cleaned seed is again allowed to dry in open sun on tarpaulin to 9.0 percent moisture.

**Seed yield :**Under good crop management, about 10-12 quintals of seed yield is obtained per hectare.

## Seed standards

The percentage of minimum physical purity of foundation and certified seeds should be 98% with a minimum of 70% of germination capacity and 9% of moisture content. The presence of inert matter should not exceed 2.0%.

# 8

# Ash gourd (*Benincasa hispida* Thumb.) Cucurbitaceae Cogn.)

**Botany:** The ash gourd(*Benincasa hispida*) belonging to the family cucurbitaceae is an annual creeping vine that can either climb structures or be allowed to spread out on the ground. This plant features large green leaves and thick stems covered with coarse hairs. The showy, golden yellow blooms appear early in the summer, and female flowers give way to round or spherical fruits. Young ash gourds are covered with a soft down that disappears with maturity. Fully matured gourds have a white, waxy coating covering the surface.

**Field preparation:**The field selected for seed production must not have been sown with ashguard in the previous season. This is done to avoid volunteer plants that cause admixture.After proper ploughing, at a spacing of 2.5 x 2 m distance take pits having 45 cm length, width and height. Ten days after that, apply 10 kg FYM and urea 30 g. Super phosphate 72 g and potash 19 g per pit. Then mix the above nutrients with soil and fill the pits and level them.

**Soil and climate:**Soil with neutral pH must be selected. Loam or clay loam soils are best suited. Higher organic matter will lead to production of vigorous seed.Seed is very sensitive to whether. Hence selecting the right season is necessary. Though ashgourd can grow through out the year, seed crop should be grown such that the seed matures in cool dry climate. This will facilitate proper ripening of fruits and reduce the pest and disease infection. Seasons are selected with this idea in mind. In Tamil Nadu the best seed crop growing seasons are Aadi and Thai Pattam *i.e.* June – July and January – February.

**Seed rate**: 2.5 kg/ha. Soak the seeds in double the quantity of water for 30 minutes and incubate for 6 days. Then five seeds may be sown in a pit at equal distance.

**Irrigation management :**After sowing pits should be irrigated with water can. Care must be taken that the soil should not be eroded and seed should not be exposed. After the seedling emergence, field should be irrigated once in a week.

**Field care:** Ten days after the germination, retain three vigorous seedlings per pit and remove two seedlings. So that it will facilitate for better growth of seedling without any competition between them.

**Controlling weed:** Maintaining the field from free of weed is more important for the crop growth. In ashgourd, one or two manual weeding is necessary before the flowering stage.

**Growth regulator spray:** Generally in curcurbitaceous vegetables, the male and female flowers borne separately. The number of female flowers decides the fruit yield. Hence, for ashguard, spraying of ethrel at 200 ppm for four times starting from four leaves stage and at weekly intervals *i.e.* 2.0 ml of ethrel in 10 lit of water is recommended. This facilitate for higher fruit yield.

**Top dressing :** Seed crop is entirely different from vegetable crop. Hence, fertilizer is also applied as split doses. Apply 22g urea/pit after 30 days after sowing as top dressing. This will facilitate the development and maturation of both fruit and seed. This leads to higher fruit bearing, higher fruit retention and quality seed yield.

**Roguing :** Roguing should be done during vegetative phase, fruit formation stage and prior to harvest.

| Vegetative phase | Fruit formation phase | Fruit harvest phase |
|---|---|---|
| During first 30-35 days, rouging is attempted based on plant characters like height of plant, leaf shape, size, surface of leaf. Plants showing heavy branching and spreading nature are removed. Plants showing variation in stem or leaf base colour are also removed. Those plants showing symptoms of yellow mosaic are also removed. | The rogues are removed based on fruit characteristics like length of fruit, size, shape and colour. | Based on fruit colour and pests and disease infection. While harvesting the workers can be taught to separate infected fruits from good ones and forward only the uninfected fruits for seed extraction. |

Field standards

| | Foundation seed | Certified seed |
|---|---|---|
| Isolation distance | 1000 m | 500 m |
| Off-types | 0.10% | 0.20% |

**Harvest :** Harvest has to be done 7-10 days after the maturity of vegetables when the fruit stalk becomes brown and dried, and as the fruits have the complete ashy coating. At the stage the seeds attain full maturity will higher vigour and viability. The harvests will be done in the different pickings in ashgourd. Here the first and last one or two harvests may be taken for vegetable. The harvest should be done from other harvests for seed extraction. Fruits confirming the genetic purity with medium to large size fruits should alone be used for seed extraction. This selection and grading procedures will increase the yield of quality seed recovery.

**Processing to improve seed quality:** After the fruit harvest and before the seed extraction, only healthy fruits of true to type and free from pest / disease infestation are to be selected for seed extraction.

## Seed extraction method

**Seed separation by cutting the fruit:** The selected fruits confirming the genetic characters alone should be used for seed extraction, and also fruits weighing less than 1.5 kg should be rejected and can be sold out in the market as vegetable. Seed extraction is easy in ashgourd. First cut the fruits into two halves by crosswise and length wise. Then remove the seed along with pulp and crash with hand in excess quantity of water. Remove the floating fraction and collect the seeds settle at the bottom.

**Seed separation:** Seed can also be extracted by acid method. Take the pulp along with seeds and Hydrochloric acid (diluted 6 times with water) at 1:1 ratio and allow it for 30 minutes with stirring. Because of this seeds will settle down. Then the floating fraction is to be removed. And then collect the seeds settled at the bottom and wash it with water for three or four times.

It is easy to dry the seeds extracted by acid method and also remove the fungal growth over the seed coat, thus seeds possess golden yellow colour and high vigour. The seed extraction by fermentation method possess poor vigour and off colour due to fungal activity.

**Seed drying:** After seed extraction, it has to be properly dried, since seeds were extracted from 100% moist condition. The extracted seeds should spread on gunny bags in a thin layer and dried under shade for 8 to 10 hours for one or two days. Then, seeds can be dried under direct sunlight between 8 to 12 noon and 3 to 5 pm. Avoid drying in between 12 noon to 3 pm, since the rays emitted from sun and the heat may affect the seed viability. While drying care must be taken to avoid clogging.

The extracted seed should not be dried directly under sun. Since seed possess high moisture it may affect the germination potential. Similar, while drying frequent stirring is more important otherwise it leads to clogging. This may results in improper drying, fungal growth and poor vigour.

**Seed cleaning and processing:** After proper drying seeds have to be processed. By removing the ill filled and small size seeds, vigour and viability is improved. For processing ashgourd seeds, first white and dull yellow ill filled seeds are to be picked out manually.

**Seed moisture:** Seed moisture is the foremost seed physical attribute that contributes for storage life. Lower he seed moisture, longer the shelf life. Short term storage can be achieved by drying the seeds to 6-7% moisture content while long term storage is possible by reducing the seed moisture even further to 6 %. Under such low moisture content, seeds have to be stored in moisture proof bags made thick polythene (700 guage).

**Seed treatment:** Prior to storage, seeds are treated with fungicide to ward off fungal pathogens. Seeds are mixed with Carbendazim 4g/kg. A novel technique called Halogen permeation treatment is also recommended in now-a-days. Calcium oxychloride, commercially know as bleaching power and powdered calcium carbonate (lime stone) is mixed in equal ratio. This mixture is added to seed at 5g / kg and stored.

**Yield:** 120-150 kg/ha

**Seed standards:** The percentage of minimum physical purity of foundation and certified seeds should be 98% with a minimum of 60% of germination capacity and 7% of moisture content. The presence of inert matter should not exceed 2.0%.

# 9

# Bitter gourd (*Momordica charantia* Linnaeus Cucurbitaceae)

Bitter gourd (*Momordica charantia*) is a widely grown vegetable variety in India. It is one of the important members of the family Cucurbitaceae. The seed production can be done throughout the year in tropical climate and during spring, summer and rainy season in subtropical climates. In Tamil Nadu September – October is suitable for seed production. In hills seed production can be done during summer season.

## Method of seed production

Bitter gourd is a self-pollinated crop with minimum cross-pollination. Seeds should be allowed to set by open-pollination in isolation. The isolation distance maintained between the fields of other varieties and the fields of the same variety not conforming to the varietal purity requirements for certification is 1000 metres for foundation and 500 metres for certified seed production.

## Seed production stages

Breeder seed → Foundation seed → Certified seed

## Land selection

The land selected should be free from volunteer plants, wild species and objectionable weeds. The land should be fertile with good drainage facility.

## Seed selection and treatment

Certified seeds should be obtained from an authorised source. Seeds should be healthy, free from disease and pest infection. Remove the broken, coloured seeds and use uniformly graded seeds. Seed rate is 2 kg/acre (4.5 kg/ha). The selected seeds should be soaked in warm water for 30 minutes before sowing. This helps in the softening of the hard seed coat of the bitter gourd seeds. To speed up the germination the seeds are kept in wet gunny bags or cloth bags for 3 – 4 days to soften the seed coat. Soaking in butter milk is also reported to promote germination. Seeds should be soaked in a solution of cow's urine (1 part cow's urine + 5 parts of water) for 30 minutes prior to the sowing. This will inhibit the seed borne diseases. Treat the seeds with *Trichoderma viride* @ 4 gms/kg of seeds. Treated seeds should be sown in the main field which is ploughed 3 – 4 times and formed with channels of 60 cm width and 2 metres of spacing. Along the channel the pits of 30 cm3 and 1 metre deep are dug at 2.5 x 2 metre spacing. Normally in summer season crop, seeds are sown in raised mounds with a spacing of 0.6 – 1.2 metre. Seeds are sown in the pits at 2 cm depth in a vertical orientation. Sow 5 seeds per pit and thin it 15 days after sowing. Allow only three seedlings per pit to grow and remove the rest. Irrigate the field before seed sowing.

### Nutrient management

Farm yard manure or compost is applied @ 10 tonnes/acre (25 tonnes/ha) before last ploughing and incorporated into the soil. In each pit, farm yard manure or compost @ 1 kg mixed with 100 gms of neem cake is applied as a basal manure. One month after sowing apply 500 gms of vermicompost per plant as top dressing. Intercultural practice should be adopted. The flower drop in the crop can be controlled by spraying Asafoetida (*Ferula assafoetida* L.) solution (125 gms of Asafoetida (*Ferula assafoetida* L.) in 1 litre of water) over the plants.

### Weed management

Weeding is most important during all the growth stages of the crop. The field should be maintained clean by frequent hand weeding. Periodical removal of objectionable weeds should be done.

### Irrigation

First irrigation is done before sowing. Subsequent irrigation should be done once a week. Irrigation during flowering and fruit setting stages are very crucial.

### Pest and disease management

| Pest/Disease | Management measures |
|---|---|
| Stem borer | Use pheromone traps @ 3 – 4/acre (8/ha), Collect and destroy the infected shoots, whole plants, fruits etc., Use *Trichoderma chilonis* @ 20,000/acre (50,000/ha) or Spray *Bacillus thuringiensis* @ 200 gms/acre (500 gms/ha). |
| Fruit fly | Managed by collection and destroying of affected fruits. |
| Leaf spot | Destroy the diseased plant debris and seed treatment using asafoetida solution (125 gms in 1 litre of water for 10 kg of seeds). |

## Roguing

Roguing should be done from early vegetative stage to flowering and fruiting stage. All the offtypes and diseased plants should be rogued off periodically. The off-types are identified based on the morphological characteristics like plant type, leaf shape, flower colour, fruit shape etc. Removal of the off-types during fruit setting stage is helpful in preventing further genetic contamination.

## Field inspection

A minimum of three field inspections should be done from vegetative to fruit maturity stage by the Seed Certification Officer. The first inspection is conducted during vegetative stage before flowering followed by the second one at flowering and fruiting stage. The final inspection should be scheduled during fruit maturity stage to determine the true characteristics of the fruits.

| Field standards | Foundation seed | Certified seed |
|---|---|---|
| Isolation distance | 1000 m | 500 m |
| Off-types | 0.10% | 0.20% |

## Harvesting

Harvesting is done once the fruits are physiologically mature. The physiological maturity of the fruits is identified by colour change from green to yellow - orange. The matured fruits are harvested by hand picking and dried until they split open and expose the shiny blood – red seeds. Seed extraction and processing. The seeds from the split opened fruits are scooped out and soaked in the water for a day to get rid of the red pulp. After this, seeds are washed repeatedly and dried. The dried seeds are graded using 16 – 64" round perforated metal sieve.

## Drying and storage

The extracted seeds should be washed and dried thoroughly before storage because of the hard seed coat in bitter gourd seeds. Before storage the moisture content of the seeds should be 7%. Under cool and dry condition the well dried seeds can be stored up to five years with good germination capacity.

## Seed standards

The percentage of minimum physical purity of the foundation and certified seeds should be 98% with a minimum of 60% of germination capacity and 7% of moisture content. The presence of inert matter should not exceed 2.0%.

# 10

# Ribbed gourd (*Luffa acutangula* Linnaeus Roxb. Cucurbitaceae)

Ribbed gourd (*Luffa acutangula (L)* Roxb.) belonging to the family Cucurbitaceae is a well known vegetable variety in India. The seed production can be done in both summer and rainy season.

## Method of seed production

Bitter gourd is a cross-pollinated crop which occurs through bees. Seeds should be allowed to set by open-pollination in isolation. The isolation distance maintained between the fields of other varieties and the fields of the same variety not conforming to the varietal purity requirements for certification is 1000 metres for foundation and 500 metres for certified seed production.

## Seed production stages

Breeder seed → Foundation seed → Certified seed

## Land selection

The land selected should be free from volunteer plants, wild species and objectionable weeds. The land should be fertile with good drainage facility.

## Seed selection and treatment

Certified seeds should be obtained from an authorised source. Seeds should be healthy free from disease and pest infection. Remove the broken, coloured seeds and use uniformly graded seeds. Seed rate is 600 gms/acre (1.5 kg/ha). The selected seeds should be soaked in warm water for 30 minutes before sowing. This helps in the softening of the seeds. Soaking in butter milk is also reported to promote germination. Seeds should be soaked in a solution of cow's urine (1 part cow's urine + 5 parts of water) for 30 minutes prior to the sowing. This will inhibit the seed borne diseases. Treat the seeds with *Trichoderma* viride @ 4 gms/kg of seeds. After 3 – 4 ploughing the main field is formed with channels of 60 cm width with a spacing of 2 metres. Along the channel, pits of

30 cm 3 diameter and 1 metre depth should be dug with a spacing of 2.5 x 2 metre. Treated seeds should be sown in the pits at 2 cm depth in vertical orientation. Sow 5 seeds per pit and thinning should be done 15 days after sowing. Only three seedlings per pit are allowed to grow and the rest are removed. Normally for summer season crop, seeds are sown in raised mounds with a spacing of 0.6–1.2 metre. The field should be irrigated before seed sowing.

## Nutrient management

Farm yard manure or compost is applied @ 10 tonnes/acre (25 tonnes/ha) before last ploughing and incorporated into the soil. In each pit, farm yard manure or compost @ 1 kg mixed with 100 gms of neem cake is applied as basal manure. One month after sowing apply 500 gms of vermicompost per plant as top dressing.

## Intercultural practice

The flower drop in the crop can be controlled by spraying asafoetida solution (125 gms of asafoetida in 1 litre of water) over the plants.

## Weed management

Weeding is most important during all the growth stages of the crop. The field should be maintained clean by frequent hand weeding. Periodical removal of objectionable weeds should be done. Irrigation First irrigation is done before sowing. Subsequent irrigation should be done once a week. Irrigation during flowering and fruit setting stages are very crucial.

## Pest and disease

| Pest/Disease | Management measures |
|---|---|
| Stem borer | Use pheromone traps @ 3–4/acre (8/ha), Collect and destroy the infected shoots, whole plants, fruits etc., Use Trichoderma chilonis @ 20,000/acre (50,000/ha) or Spray *Bacillus thuringiensis* @ 200 gms/acre (500 gms/ha). |
| Army worm | Use light traps @ 2/acre (5/ha) and spray five leaf extract or ginger, garlic and chilli extract @ 1 litre/tank. |

## Roguing

Roguing should be done from early vegetative stage to flowering and fruiting stage. All the offtypes and diseased plants should be rogued off periodically. The off-types are identified based on the morphological characteristics like plant type, leaf shape, flower colour, fruit shape etc. Removal of the off-types during fruit setting stage is helpful in preventing further genetic contamination. The maximum percentage of off-types permitted at final inspection is 0.10% for foundation seed production and 0.20% for certified seed production.

## Field inspection

A minimum of three field inspections should be done from the vegetative to fruit maturity stage by the Seed Certification Officer. The first inspection is conducted during vegetative stage before flowering followed by the second one at flowering and fruiting stage. The final inspection should be scheduled during the fruit maturity stage and prior to harvest.

## Field standards

| Field standards | Foundation seed | Certified seed |
|---|---|---|
| Isolation distance | 1000 m | 500 m |
| Off-types | 0.10% | 0.20% |

## Harvesting

Harvesting is done once the fruits are physiologically mature. The physiologically mature fruits will dry on the vine with cracked skin. The matured fruits should be harvested by hand picking and dried further.

## Seed extraction and processing

The harvested fruits are dried under the sun for 2 -3 days. In well dried fruit shell, the seeds will rattle inside. Seed extraction in ribbed gourd is very simple and easy. The harvested, dried fruits should be cut open at one end and shaken. The seeds held by dry fibre will come out. These seeds can be collected and conserved. No further cleaning is required.

## Drying and storage

The extracted seeds should be dried to attain a moisture level of 7% before storage. The dried seeds are graded using 16 – 64" round perforated metal sieve. Under dry, cool and dark storage conditions the well dried seeds can be stored up to five years.

## Seed standards

The percentage of minimum physical purity of foundation and certified seeds should be 98% with a minimum of 60% of germination capacity and 7% of moisture content. The presence of inert matter should not exceed 2.0%.

# 11

# Snake gourd (*Trichosanthes anguina* Linnaeub Cucurbitaceas)

**Botany:**Snake gourd belonging to the family Cucurbitaceae native to Southeastern Asia and Australia but cultivated throughout the world for its curved and oddly shaped fruits that appear like snakes hanging on the supports or ground. This subtropical plant grows very fast in warm climates and produces lots of fruits for a long time. It is suitable for growing for home garden and fresh market.

**Climate and soil:** Snake gourd is adapted to wide variety of soil and climatic conditions. It requires a minimum temperature of 18°C during early growth, but optimal temperatures are in the range of 24–27°C. Snake gourd tolerates a wide range of soil but prefers a well drained sandy loam soil that is rich in organic matter. The optimum soil pH is 6.0–6.7, but plants tolerate alkaline soils up to pH

**Varieties:** CO1,CO2, PKM 1, PLR 1 and PLR 2.

**Season and sowing:** July and January. Sow the seeds (3 seeds/pit) treated with Trichoderma viride @ 4 g/kg or Pseudomonas fluorescens @ 10 g/kg or carbendazim @ 2 g/kg and thin the seedlings to two per pit after 15 days.

**Seed rate:** 2.5 kg/ha.

**Preparation of field**: Plough the field to fine tilth. Dig pits of size 30 cm x 30 cm x 30 cm at 2.5 x 2 m spacing and form basins.

**Pre-sowing treatment:** The seeds are pre-germinated by incubating the seed for 4 days in between gunnies after soaking in double the volume of water for 4 h. Then, pre-sprouted seeds are separated and used for sowing.

**Irrigation:** Irrigate the basins before dibbling the seeds and thereafter once in a week.

**Application of fertilizers:** Apply 10 kg of FYM, 100 g of NPK 6:12:12 mixture as basal dose per pit and N @ 10 g pit 30 days after sowing. Apply *Azospirillum* and Phosphobacteria @ 2 kg/ha and *Pseudomonas* 2.5 kg/ha along with FYM 50 kg and neem cake @ 100 kg before last ploughing.

**Foliar application:** Spraying maleic hydrazide @ 400 ppm or ethrel at 250 ppm at 2 leaf and 5 leaf stages enhances female flower production.

## Plant protection

### Pests

**Leaf beetles and leaf caterpillars:** Spray Dichlorvos 76% EC 6.5 ml/10 lit or Trichlorofon 50% EC 1.0 ml/l.

### Fruit fly

- Collect the damaged fruits and destroy.
- The fly population is low in hot day conditions and it is peak in rainy season. Hence, the sowing time may be adjusted accordingly.
- Expose the pupae by ploughing.
- Use 20 x 15 cm poly bags fish meal traps with 5 g of fish meal + 1 ml of Dichlorvos in cotton @ 50 traps/ha. Fish meal and cotton are to be removed once in 20 and 7 days respectively.
- Neem oil @ 3.0 % as foliar spray as need based
- For management of Aphid vector, spray Imidachloprid @ 0.5 ml/lit along with sufficient quantity of stickers like Teepol, triton X100, apsa etc., for better adhesion and coverage.
- Do not use copper and sulphur dust. These are phytotoxic.

### Diseases

**Powdery mildew:** Spray Dinocap 1 ml/l or Carbendazim 0.5 g/l.

**Downy mildew:** Spray Mancozeb or Chlorothalonil 2 g/l twice at 10 days interval.

## Field standards

| | Foundation seed | Certified seed |
|---|---|---|
| Isolation distance | 1000 m | 500 m |
| Off-types | 0.10% | 0.20% |

**Harvest :** Change of fruit colour in any part or 1/3 of fruit tip to yellow to red for seed extraction

**Seed extraction:** Fruits are cut and seeds with pulp are scooped out and seeds are separated by washing with water and are dried to 7- 8 % moisture content.

**Grading:**The immature seeds can be removed as water floaters during wet extraction. Seeds are graded with 16/64" round hole sieve or BSS 4 x 4 (6.2 mm) for homogenizing the lot. Seed yield: 300-350 kg/ha

**Storage:**Seeds dried to 7-8% moisture content and dry dressed with carbendazim 50 % WP @ 2 g / kg seed or halogen formulation (Bleaching powder + $CaCO_3$ + Arappu *Albizia amara* (Roxb.) Boiv leaf powder @5:4:1) @ 3 g/kg seed and stored in cloth bag upto 10 months and more than 18 months in moisture vapour proof containers.

**Yield:** 220-250 kg/ha

**Seed standards :**The percentage of minimum physical purity of foundation and certified seeds should be 98% with a minimum of 60% of germination capacity and 7% of moisture content. The presence of inert matter should not exceed 2.0%

# 12

# Bottle gourd (*Lagenaria siceraria* (Molina) Standley Cucurbitaceae)

**Botany:** Bottle gourds belonging to the family cucurbitaceae is an annual vine having white flowers and smooth, large, hard-shelled gourds. Brownish seeds are numerous in a whitish green pulp. Each seed is somewhat rectangular in shape with grooved notches near the attached end. Grown most often in warmer climates, this squash grows from 6 to 36 inches long and 3 to 12 inches in diameter.

Seeds leaves, flowers and young stems are all edible. Helpful in shedding extra calories. It contains higher concentrations of Dietary Fiber, Vitamin A, Vitamin C, Vitamin K, Vitamin B6, Folate, Potassium, Manganese, Protein, Vitamin E, Thiamin, Riboflavin, Pantothenic Acid, Calcium, Iron, Magnesium, Phosphorus and Selenium. A rich source of minerals and vitamins, bottle gourds contains many healing and medicinal properties. The cooked vegetable is not only easy to digest but also contains cooling, calming (or sedative), diuretic properties. It contains low calories also has iron, Vitamin C and B complex. Regular consumption of this vegetable provides relief to people suffering with digestive problems, diabetics and convalescents.

**Soil:** Sandy loams rich in organic matter with good drainage and a pH range from 6.5 to 7.5.

**Season and sowing:** July and January. Sow the seeds (3 seeds/pit) treated with *Trichoderma viride* @ 4 g or *Pseudomonas fluorescens* 10 g or Carbendazim 2 g/kg of seeds and thin the seedlings to two per pit after 15 days.

**Varieties:** Pusa Summer Prolific Long, Pusa Summer Prolific Round, Pusa Manjari, Pusa Megdoot and Arka Bahar.

**Seed rate:** 3 kg/ha.

**Spacing:** Pit size : 45 x 45 cm at a distance of 2.5 x 2 m Foliar application: Spraying Maleic hydrazide @ 400 ppm or ethrel at 250 ppm at 2 leaf stage and 5 leaf stage enhances the female flower production and seed yield.

**Preparation of field**: Plough the field to fine tilth. Dig pits of 30 cm x 30 cm x 30 cm size at 2.5 x 2 m spacing.

**Irrigation:** Irrigate the field before dibbling the seeds and thereafter once a week.

**Manures and fertilizers:**Apply 10 kg of FYM (20 t/ha) and 100 g of NPK 6:12:12 mixture as basal and 10 g of N per pit 30 days after sowing. Apply *Azospirillum* and Phosphobacteria 2 kg/ha and *Pseudomonas* 2.5 kg/ha along with FYM 50 kg and neem cake @ 100 kg before last ploughing.

**After cultivation:** Hoe and weed thrice.

## Plant protection

### Pests

Mites: Spray Dicofol 18.5 % SC @ 2.5 ml per litre of water Aphid: Spray Imidachloprid @ 0.5 ml/lit along with sufficient quantity of stickers like Teepol, triton X100, apsa etc., for better adhesion and coverage.

### Diseases

**Powdery mildew:** Spray Dinocap 1 ml/l. or Carbendazim 0.5 g/l or Tridemorph 1ml/l.

**Downy mildew:** Spray Mancozeb or Chlorothalonil 2 g/l. twice at 10 days interval.

## Field standards

| | Foundation seed | Certified seed |
|---|---|---|
| Isolation distance | 1000 m | 500 m |
| Off-types | 0.10% | 0.20% |

**Harvest:** Fruits attain physiological maturity at 65 days after anthesis (the skin of the fruit become woody rough and turn dull in colour). Fruits weighing less than 50 g should be rejected as it contains higher percentage of immature seeds.

**Seed extraction:** Matured fruits are cut vertically and seeds are scooped and cleaned.

**Grading:** Seeds are graded with 16/64" round perforated metal sieves or BSS 4 x 4 wire mesh sieve (6.2 mm).

**Seed yield:** 250 kg/ha

**Storage:** Seeds dried to 8 % moisture content and treated with carbendazim 50 % WP @ 2 g / kg seed or halogen formulation (Bleaching powder + $CaCO_3$ + Arappu *Albizia amasa* (Roxb.) Boiv. leaf powder @ 5:4:1) @ 3 g/kg seed can be stored upto 1 year in cloth bag and 2 years in moisture vapour proof containers.

**Seed standards :**The percentage of minimum physical purity of foundation and certified seeds should be 98% with a minimum of 60% of germination capacity and 7% of moisture content. The presence of inert matter should not exceed 2.0%

# 13

# Pumpkin (*Cucurbita maxima* Duch. Cucurbitaceae)

Pumpkin (*Cucurbita maxima* Duch.) is a well known and widely cultivated vegetable variety of the family Cucurbitaceae. The seed production can be done in summer season (February - March) and in rainy season (April - May). In Tamil Nadu, sowing during July – August is best for seed production.

## Method of seed production

Pumpkin is a cross-pollinated crop and self-pollination occurs to the extent of 5%. Seeds should be allowed to set by cross-pollination in isolation. The isolation distance maintained between the fields of other varieties and the fields of the same variety not conforming to the varietal purity requirements for certification is 1000 metres for foundation and 500 metres for certified seed production.

## Seed production stages

Breeder seed → Foundation seed → Certified seed

## Land selection

The land selected should be free from volunteer plants, wild species and objectionable weeds. The land should be fertile with good drainage facility.

## Seed selection and treatment

Certified seeds should be obtained from an authorised source. Seeds should be healthy and free from disease and pest infection. Remove the broken, coloured seeds and use uniformly graded seeds.

Seed rate is 400 gms/acre (1 kg/ha). Soak the seeds in double the quantity of water for 4 hours and bundle them in a wet cotton cloth for five days to ensure the uniform emergence of the seeds in the field. Seeds should be soaked in a solution of cow's urine (1 part cow's urine + 5 parts of water) for 30 minutes prior to the sowing. This will inhibit the seed borne diseases. Treat the seeds with *Trichoderma viride* @ 4 gms/kg of seeds. The treated seeds are sown directly in the field in raised beds or furrows or in pits @ 2 seeds per hill at 4 – 5 cm distance. For sowing in raised beds and furrows the row to row spacing of 2 – 2.5 metres and plant to plant spacing of 100 – 150 cm should be followed. For pit sowing, pits of 60 x 60 x 60 cm should be dug. After the preparation of the main field, 45 cm wide and 25 – 30 cm deep channels are formed with a spacing of 3 – 4.5 metre. Sowing in 'channel and hill' method is very effective. Seeds are sown in the pits at 2 cm depth in vertical orientation. The field should be irrigated before seed sowing.

### Nutrient management

Farm yard manure or compost is applied @ 10 tonnes/acre (25 tonnes/ha) before last ploughing and incorporated into the soil. In each pit, farm yard manure or compost @ 1 kg mixed with 100 gms of neem cake is applied as a basal manure. One month after sowing, apply 500 gms of vermicompost per plant as top dressing.

### Intercultural practice

The flower drop in the crop can be controlled by spraying asafoetida solution (125 gms of Asafoetida (*Ferula asafoetida* L.) in 1 litre of water) over the plants.

### Weed management

Weeding is most important during all growth stages of the crop. The first weeding can be done 15 – 20 days after seed sowing. Frequent weeding should be done to keep the field clean. Periodical removal of objectionable weeds should be done.

### Irrigation

Regular irrigation is a must for optimum yield. First irrigation is done at the time of sowing followed by the life irrigation at 3 days after seed sowing. Subsequent irrigation should be done at an interval of 5 – 7 days. Irrigation during flowering and fruit setting stages are very crucial. But, excessive irrigation during fruit maturity should be avoided for better storage life.

### Pest and disease management

| Pest / Disease | Management |
|---|---|
| Powdery mildew | Spray butter milk extract (two parts of water in one part of curd) @ 1 litre / tank or *Eucalyptus* leaf extract @ 10%. |
| Downy mildew | Remove infected seedlings at the time of thinning and plants from time to time. |
| Fruit fly | Treat seeds with *Trichoderma viride* @ 4 gms/kg of seeds and remove and destroy the affected and decayed fruits. |

## Roguing

Roguing should be done from early vegetative stage to flowering and fruiting stage. All the off-types and diseased plants should be rogued off periodically. The off-types are identified based on the morphological characteristics like growth pattern, flowering, fruit shape etc. These off-types should be rogued off immediately to conserve the genetic purity of the seeds. The maximum percentage of off-types permitted at the final inspection is 0.10% for foundation seed production and 0.20% for certified seed production.

## Field inspection

A minimum of three field inspections should be done from vegetative to fruit maturity stage by the Seed Certification Officer. The first inspection is conducted during vegetative stage before flowering followed by the second one at flowering and fruiting stage. The final inspection should be scheduled during fruit maturity stage and prior to harvest.

| Field standards | Foundation seed | Certified seed |
|---|---|---|
| Isolation distance | 1000 m | 500 m |
| Off-types | 0.10% | 0.20% |

## Harvesting

Harvesting is done once the fruits are physiologically mature. The physiological maturity can be identified by colour change from green to yellow and drying of the fruit stalks. The matured fruits should be harvested by hand picking and stored for few weeks for further maturation of seeds.

## Seed extraction and processing

The seeds are scooped from the matured fruit and washed thoroughly to remove the pulp around them. After washing the seeds are dried under the shade to attain safe moisture content. The dried seeds are graded using 16/64" round perforated metal sieve.

## Drying and storage

The extracted seeds should be put in paper envelopes and hung out for further drying for a week to attain a moisture level of 7% before storage. Under suitable storage conditions of dry environment with even temperatures the seeds can be stored to an extent of 3 – 10 years.

## Seed standards

The percentage of minimum physical purity of foundation and certified seeds should be 98% with a minimum of 60% of germination capacity and 7% of moisture content. The presence of inert matter should not exceed 2.0%.

# 14

# Onion (*Allium cepa* Linnaeus Alliaceae)

Onion is one of the oldest cultivated plant species. Onion forms the bulb in the first year and the seed in the second year. The leaves arise from a shortened crown stem. The sheaths of the older or outermost leaves enclose the younger ones. The basal portion of the leaves encircles the stem, and thickens to form the bulb. The stem elongates during the second year forming the flower stalk. The onion has a fibrous root system extending to a depth of over one meter, but most of them are found in upper surface about one-half meter area.

## Floral Biology and Nature of Pollination

Flowers are borne in simple umbels at the apex of a floral stem which is hollow and round in cross section and somewhat swollen at the middle or near the base. Most onion varieties produce seed stalks over one meter height. The number of seed stalks per plant may vary from 1 to 20 or more depending on the variety, size of mother bulb and time of plating. Before expanding, the umbel is enclosed within a papery spathe consisting of 2 or 3 bracts which are split open by the pressure of the developing flower buds. The number of flowers per umbel varies considerably, from 50 to over 2000. The differentiation of the flower begins in the late winter. Temperatures around 20-22°C favour the vegetative growth while temperatures around 12-13°C are conducive to seedstalk formation. Also short day conditions are favourable to seed production.

White or bluish flowers have an outer and an inner whorl of stamens of three each. The anthers of inner stamens dehisce first. The pistil has a three-celled ovary with two ovules in each. The style is about one mm in length when the flowers open first. It is not receptive until it elongates to a length of about 5 mm, which requires 1 or 2 days after all the anthers have dehisced. Opening of flowers usually continues for a period of two weeks or more and onion plant may be in bloom for over 30 days. The fruit is a three-lobed, three-celled capsule, each locule containing 1 or 2 black seeds at maturity.

## Climate and seasons

Onion flowering is a thermo-sensitive phenomenon, short day tropical types flower under low chilling (25°C day and 10–15°C night), long day temperate types requires high chilling (0–50°C), Oct.-Nov. planting is the best time for tropical types. As per recommended package of practices, the mother bulbs of Rabi crop should be produced and has to be stored up to October. Bulbs must be stored in well-ventilated storage structure with temperature 25-30°C and RH 65-71%. Plants developed from bulbs stored at optimum temperature usually flower and mature early and produce higher seed yield.

## Varieties

Bellkary Red, Rampur local, Pusa white, Kalyanpur, Punja 48, Pusa red, Pusa Madhvi, Arka, Nikertan, Arka Kalyani

## Method of seed production

### a) Seed to seed method

Seedlings after transplanting are allowed to over winter in the field. The bulbs are not lifted and allowed to flower in the same field. This method is not preferred because it doesn't allow the examination of true to typeness of the onion bulb and rouging of off types and diseased and multicentre bulbs. The genetic purity of the seeds produced from such method is usually poor, though the cost of seed production is less compared to bulb to seed method but this method is not preferred.

### b) Bulb to seed method

In the first season the onion bulbs are produce which after maturity are harvested and bulbs which are true to type are sorted and stored. The *rabi* and late *kharif* varities are stored for next season and *kharif* varities are given rest for 15-30 days. The stored bulbs are planted in the next season to produce seeds. Top 1/3 portion of the bulb is cut to examine the number of axis of growing center, preferably single center bulbs should be chosen and soaked in 1g/litre bavistin & corbosulphon 1m/litre. This method is widely practised it allows us to examine the bulb characteristics and rouging of undesirable bulbs hence genetic purity is high in this method of seed production

## Selection of planting material

Bulbs should be true to type in colour and shape, medium to large in size with preferably single centre

## Planting Spacing

| Spacing (cm) | Av. Bulb Weight(g) | Quantity of bulbs (q/ ha) |
|---|---|---|
| 45 x 30 | 50 60 70 80 | 26 31 36 41 |
| 60 x 20 (on drip) | 50 60 70 80 | 30 35 40 45 |

## Field standards

| Contaminants | Minimum Distance (meters) | | | |
|---|---|---|---|---|
| | Mother bulb Production | | Seed Poduction Stage | |
| | Foundation | Certified | Foundation | Certified |
| Fields of other variety | 5 | 5 | 1000 | 500 |
| Fields of the same variety conforming to varietal purity requirements for certification | 5 | 5 | 1000 | 500 |

## Fertilizers

FYM @ 25 tons/ha, NPK @ 100:50:50 kg/ha apply 50:50:50 kg/ha NPK at the time of planting & remaining nitrogen in 2 splits, one at 30 days and 2nd at 45-60 days after planting. Give 1% spray of Polyfeed (19:19:19, NPK) at 30 & 60 days after planting & one spray of multi K (0:0:50) after 60 days of planting

## Roguing

Plot should be visited regularly. Yellow and lanky plants should be removed before flowering; plants with differential umbel height should be removed before opening of flowers; plants affected by aster yellow and stemphyllium blight should be removed before seed harvest.

## Weed Management

Spray Goal (Oxy fluorfen) @ 1.5 ml/L after planting of bulbs and one weeding after 45 to 60 day after planting should be done.

## Pest and disease control

| Pest and disease | Control measures |
|---|---|
| Onion thrips | Dust with 5% HCH dust at 25 to 30 kg per hectare, orspray malathion 50 EC at 600 to 700 ml per hectare or Endosulpfan 35 EC at 600 to 700 ml per hectare. |
| Onion maggot | Spray Sevimol |
| Damping-off | Use treated seed.In cases of seedlings mortality, drench nursery with 0.3 per cent captan, or Thiram, or Dithane Z-78 at weekly intervals. |
| Purple blotch | Spray with copper fungicides such as Blitox 50 at 0.2 per cent. |

## Drying and Threshing

Dry umbels in open sun, threshing of seed can be done by rolling, threshing machine or combines. The seed should be dried in open sun till 6 to 7 per cent moisture level is attained. Seed should be packed in 400 gauge poly bags after perfect drying. Seeds should be stored at Room temperature -30-35°C - 15-18 months Cold Storage - 15 °C with 30–40%RH - 3-4 Years

## Average Seed yield

500 – 800 kg seed/ha, in best management and climatic conditions 1000 to 1200 kg seed can be obtained

## Seed standards

The percentage of minimum physical purity of foundation and certified seeds should be 98% with a minimum of 70% of germination capacity and 8% of moisture content. The presence of inert matter should not exceed 2.0%.

# 15

# Amaranth (*Amaranthus tricolor* Linnaeus Amaranthaceae / *A.cruentus* Linnaeus / *A.blitum* var. *oleracea Duthie*)

Amaranth is an annual herb with erect growth and scarce to profuse branching habit. Flowers are borne terminally and in axils of leaves in clusters. Basic unit of inflorescence is called as glomerule. Flowers are small, unisexual and monoecious. Proportion of male and female flowers varies in an inflorescence. Each glomerule consists of a staminate flower and a number of pistillate flowers. The extent of cross pollination is governed by proportion of male and female flowers in an inflorescence and position of inflorescence in plant

### Climate and Soil

Amaranthus is widely distributed in both tropical and sub tropical regions. Leaf amaranth is a warm season crop adapted to hot humid climatic conditions. It is grown throughout the year in tropics and in autumn, spring and summer seasons in temperate regions. Most of leaf types are day neutral in habit but differ in their day length requirements and respond differently to changes in photo and thermoperiodism.

Grain types, *A. caudatus, A. cruentus* and *A. edulis* are short day species while A. hypochondriacus is day neutral. Amaranth comes up well in well drained loamy soil rich in organic matter. The ideal is pH is 5.5-7.5 but there are types which can come up in soils with pH as high as 10.0. Red amaranth requires bright sunlight for colour development.

### Varieties

Cultivated leaf amaranth varieties and cultivars differ in size, shape and colour of leaves and stem, position of inflorescence etc. and belong to different species. A brief description of improved varieties developed by different Research Institutes is given below: 1) Tamil Nadu Agricultural University, Coimbatore-3

| | |
|---|---|
| CO. 1 (*A. dubius*) | This tetraploid variety was developed by selection from ("local germplasm. Stem and leaves are dark green; leafstem ratio is 2.0; inflorescence terminal and axillary; lacks initial vigour but makes rapid growth after 30 days; suitable for late harvest; resistant to *Rhizoctonia* leaf blight; green yield 8.0 t/ha; seed yield 1.5 t/ha. |
| CO.2 (*A. tricolor*) | Stem and leaves green, leaves lanceolate and slightly elongate, leaf-stem ratio 1.8; suited for early harvest; yield 10.78 t/ha. |
| CO.3 (*A. tristis*) | This is specifically suited for clipping of tender greens and is locally known as 'Araikeera' in Tamil. Leaves are small and green; stem is slender and tender. First clipping is possible in 20 days after sowing. Nearly 10 clippings can be taken over a period of 90 days. Due to very high leafstem ratio, cooking quality and taste are excellent. Special care is required in land preparation for the variety. |
| CO.4 *A. hypochondriacus*) | This grain type makes rapid vegetative growth within a period of 20-25 days. Plants are dwarf; grain yield 2.0-2.5 t/ha in 80-90 days. |
| CO.5 (*A. tricolor*) | Leaves double coloured with Green and pink and is free from fibre. It gives a rosette growth in early stages and first harvest is possible in 25 days; yield 40 t/ha in 55 days. |

**Sirukeerai** (*A. polygonoides*) is a traditional cultivar in Tamil Nadu, suited for uprooting at 25 days after sowing; leaves are small, ovate with blunt bifurcated tip and have long petiole; collar region is dark pink and at leaf axil a miniature branch initiates.

### IARI, New Delhi

Pusa Chotti Chaulai (*A. blitum*): Plants dwarf with succulent, small and green leaves; responds well to cutting.

Pusa Badi Chaulai (*A. tricolor*): Plants tall and stem thick with large green leaves; responds to cutting.

Pusa Kirti (*A. blitum*): Green leaved variety with green and thick stem; leaf lamina broad ovate; ready for harvest in 30-35 days and extends up to 70-85 days; yield 55 t/ha; specifically suited for summer.

Pusa Kiran (*A. tricolor*): This is developed by natural crossing between A. tricolor and A. tristis and has more characteristics of A. tricolor. Leaves are glossy green with broad ovate lamina; leaf-stem ratio is 1.0:4.6; yield 35 t/ha in 70-75 days; suited for *kharif* season.

Pusa Lal Chaulai (*A. tricolor*): Upper surface of leaves are deep red and lower surface purplish red; yield 45-49 t/ha in 4 harvests.

### IIHR, Bangalore

Arka Suguana (*A. tricolor*): A multicut variety with broad green leaves. First picking starts in 24 days after sowing and continue up to 90 days. Moderately resistant to white rust. Yield 17- 18 t/ha.

Ark Arunima (*A. tricolor*): *A multicult* variety with broad dark purple leaves. First picking starts in 30 days after sowing and two subsequent cuttings at 10-12 days interval. Yield 27 t/ha.

### Land preparation and sowing

Amaranth is harvested by pulling out and by frequent clippings (multicut). Cultivation practices differ according to method of harvest, duration, growth pattern of variety, etc. Land is prepared to a fine tilth by thorough ploughing and harrowing. Well decomposed and powdered organic matter @ 20-25 t/ha is incorporated with the soil at the time of final ploughing.

### Isolation Requirement

A minimum isolation distance of 400 meters and 200 meters for foundation seed and certified seed class respectively is required.

### Manures and fertilizers

Amaranth is a heavy feeder and high yielding crop. 20-25 tonnes of FYM and 50:25:20 kg NPK / ha are recommended as basal dose. Under pulling out method, 20 kg N should be top dressed twice during subsequent pulling out of seedlings. For clipping varieties, a still higher dose of 75:25:25 is advisable. Apply N after every clipping or cutting. Foliar spray of 1% urea or diluted cow urine at every harvest is good for promoting further growth and for high yield.

### Majaor Pests and diseases

| Diseases | Control measures |
|---|---|
| Leaf blightand white rust | Sow resistant green amaranth variety, CO-1 during rainy season. Avoid splash irrigationSpray Mancozeb @ 4g/1 of cow dung supernatant as fine droplets.Cover plants thoroughly so that spray solution reaches under surfaces of leaves also. |

### Roguing

A minimum of two inspections during vegetative stage followed by flowering stage is to be carried out. Rouge out off types and wild *Amaranthus spp.* from seed fields prior to flowering and during flowering.

### Field inspection

A minimum of two inspections should be done from vegetative stage to flowering stage by the Seed Certification Officer. The first inspection is done before flowering followed by the second during flowering stage to determine isolation, off-types, volunteer plants and diseased plants and to estimate the yield.

| Field standards | Foundation seed | Certified seed |
|---|---|---|
| Isolation distance | 400 m | 200 m |
| Off-types | 0.10% | 0.20% |
| Objectionable weed seeds | 0.01% | 0.02% |

## Cutting

Periodic cuttings may be taken as usual. For seed crop the last one or two cuttings are not taken.

### Harvesting and threshing of seeds

The seeds reach the physiological maturity in 25 days after flowering. Harvesting takes place soon after the maturation of seeds. The physiological maturation of the glumes and seeds are identified by colour change from green to brown and green to shiny black, respectively. Seed heads (glumes) nearing maturation should be harvested now and then, since seeds tend to drop from the fully matured glumes.

### Threshing and processing

Harvested seed heads are dried under the sun light to attain a moisture level of 15%. After this, using a pliable bamboo stick the glumes are beaten to shed the seeds. The separated seeds are then cleaned to remove the debris. Winnowing is avoided since the seeds are very small and less in weight. To separate the debris from the seeds, the seeds should be heaped in a bowl and tossed. The debris will collect at the top and can be blown away. Then the seeds are graded using BSS 22 x 22 wire mesh sieve.Seeds are finally dried to 7% moisture and stored.

**Seed yield:** 2 3 quintals per hectare

### Seed standards

The percentage of minimum physical purity of foundation and certified seeds should be 95% with a minimum of 70% of germination capacity and 8% of moisture content. The presence of inert matter should not exceed 5.0%.

# 16

# Moringa (*Moringa oleifera* Linnarck Moringaceae)

**Botany**

The plant is highly cross pollinated due to heteromorphism and bees being the major pollinators.

**Varieties:** PKM 1 & PKM 2

**Method of seed production :** Seed to Seed.

**Land requirement :** Uproot all the wild and non specific moringa trees if any near the field within a radius of 1500ft. Good drainage facility is must for moringa.

**Isolation requirement :** A distance of at least 500 M is necessary for seed production.

**Season :** Early season (July  August), Mid season (September), Late season (October -November). Best season is September  October, so that flowering starts during summer which facilities better pollination and seed set.

**Pit preparation :** Plough the field twice for good filth. Form a pit with 45 x 45 x 45 cm size, one week prior to sowing / planting. Spacing between the pits must be 2 ½ x 2½ m. Apply 15 kg FYM or compost per pit with field soil and fill the pit.

**Sowing :** Seeds are readily germinable (non dormant). However, soaking the seeds in water, overnight will hasten the germination processes. Place two seeds per pit in the sowing depth of -2.5 cm. normally, the seeds germinate within 7 days.

Simultaneously fill the FYM + soil mixtures in poly bags and place one seed per bag and keep it as reserve for gap filling. At least 50 seedlings are needed for gap filling.

**Seed rate:** 450 g/ha

**Irrigation :** Irrigate the field once in a week upto 3 months and once in a 10 days thereafter. Water stagnation should be avoided as it leads to flower drop. During pod development, irrigate once in 5 days for better pod development.

**After cultivation:** Pinching is done when the plant is 90-100 an height for better establishment of side branches. 2 pinching may be done further, at 20-25 days intervals.

**Flowering :** Flowering starts 5-6 months after sowing. Pod and seeds take 70 - 90 days to develop. During flowering, irrigation should be restricted to avoid flower dropping and liberal irrigation should be given during pod development. Apply micronutrient and spraying of Planofix (NAA) @ 4ml/L to arrest the flower shedding.

**Manuring :** Apply 100 g urea, 100 g Super Phosphate and 50 g MOP per pit, 3 months after sowing again apply 100 g urea per pit at the time of flowering.

**Pest and disease management:** The major pests are fruit fly, aphid and jassids which can be controlled by Carbaryl spray @ 2g/L or any systemic insecticides. The diseases like root rot can be controlled by root drenching with Copper oxychloride @ 2g/L.

**Roguing:** Based on the plant stem characters, during early stage, the rogues should be completely pulled out and gap may be filled. During pod development and maturity stages, based on pod character the roguing should be done for example the pods with more than 70 cm and cylindrical shape alone should be harvested in case of PKM1. Pods with tri-faced shape should be rejected.

**Harvesting and processing:** The change of pod colour to brown is the maturity index, at that stage the seed colour will be black. 3-4 pickings may be done. After harvest, dry the pods under the sun for 2 days and extract the seeds manually by split open the pod. Sundry the seeds during morning and evening hours. Dry the seeds to 7-8°h moisture content level. Separate the small, ill filled, brown coloured and damaged seeds manually. Treat the seeds with Carbendazim @ 4 g per kg of seed and pack in cloth / polythene bags. Normally seeds are viable for one year.

**Seed yield:** Approximately 150 pods per tree with 15 seeds per pod *i.e.*, 500g seed per tree and approximately 250 kg /ha

**Ratooning**: One ratooning can be allowed for certified seed production. However, more ratooning can be practiced for truthfully labeled seed production. Cut the trees at a height of 90 cm and follow the same cultural operations as followed for main crop. Apply 25 kg FYM/pit for ratoon crop and irrigate the field immediately after ratooning.

## Seed standards

The percentage of minimum physical purity of foundation and certified seeds should be 96% with a minimum of 70% of germination capacity and 8% of moisture content.

# 17

# Elephant yam (*Amorphophallus companulatus* Blume Araceae)

Elephant foot yam is an underground stem tuber belonging to the family Araceae, one of the most popular tuber crops, extensively used as a favourite vegetable by millions of people in India. It has both nutritional and medicinal value and is usually consumed as cooked vegetable. It has high dry matter production capability per unit area than most of the other vegetables. Elephant foot yam is a remunerative and profitable stem tuber crop. The crop is gaining popularity due to its shade tolerance, easiness in cultivation, high productivity, less incidence of pests and diseases, steady demand and reasonably good price.

Tubers are mainly used as vegetable after thorough cooking. Chips are made of starch-rich tubers. Tender stem and leaves are also used for vegetable purpose. Tubers contain 18.0% starch, I-5% protein and up to 2% fat. Leaves contain 2-3% protein, 3% carbohydrates and 4-7% crude fibre. Tubers and leaves are quite acrid due to high content of oxalates. Acridity is usually removed by boiling fairly for a long time. Cultivation of elephant foot yam is limited to India, Philippines, Sri Lanka and South East Asia.

**Varieties :**Gajendra and Sree Padma are the popular cultivars.

**Soil :**A rich red-loamy soil with a pH range of 5.5-7.0 is preferred. It is a tropical and subtropical crop. It requires well distributed rainfall with humid and warm weather during vegetative phase and cool and dry weather during the corm development period.

**Season and planting** :It undergoes a dormancy period of 45 to 60 days. Traditionally farmers take advantage of the dormancy period by planting during February-March so that the setts would sprout with the pre-monsoon showers. April – May is the planting season. The tuber is cut into 750-1000g small bits in such a way that each bit has atleast a small portion of the ring around each bud. Whole corms of 500 g size can also be used as a planting material. Use of cormels and minisett transplants of 100 g size as planting material at a closer spacing of 45 x 30 cm is also suggested. There are also projections with tender buds called “Arumbu”. These are removed before planting as they do not give vigorous growth. An ordinary sized yam gives about 6 to 8 bits for planting. The cut pieces are dipped in cow dung solution to prevent evaporation of moisture from cut surface. In some places, the small round daughter corms are also planted. The cut pieces are planted in beds at 45 cm x 90 cm spacing or pit of 60 x 60 x 45 cm size is dug and planted. The pit should be filled with top soil and farm yard manure (2kg/pit) prior to planting. The pieces are planted in such a way that the sprouting region (the ring) is kept above the soil. About 3500 kg of corms will be required to plant one hectare. Sprouting takes place in about a month.

**Preparation of field** :The land is brought to fine tilth and form beds of convenient size.

**Planting :**The cut pieces are planted in beds at 45 cm x 90 cm spacing. The pieces are planted in such a way that the sprouting region (the ring) is kept above the soil. Sprouting takes place in about a month.

**Intercropping :**Vegetable cowpea var. CO2 is recommended as suitable intercrop in elephant foot yam. It can be intercropped profitably in coconut, arecanut, rubber, banana and robusta coffee plantations at a spacing of 90 x 90 cm. Half quantity of FYM (12.5 t/ha) and one third of NPK (27:20:33) will be sufficient for the intercrop.

**Irrigation** :It is mostly raised as a rainfed crop. However, irrigation is required when monsoon fails, where it is grown on a large scale. Water stagnation is harmful to the crop. Wherever irrigation facility is available, irrigation can be given once a week.

**Application of fertilizers** :Apply 25 tonnes of FYM/ha during last ploughing. The recommended dose of NPK/ha is 80:60:100 kg. Apply 40:60:50 kg NPK/ ha at 45 days after planting along with weeding and intercultural operations. Top dress with 40:50 N and K one month later along with shallow intercultural operations.

**After cultivation** :Weeding and earthing up as and when necessary.

## Plant protection

### Disease

**Leaf spot :**Leaf spot disease can be controlled by spraying Mancozeb at 2g/lit.

**Collar rot** :The disease is caused by a soil borne fungus *Schlerotium rolfsii.* Water logging, poor drainage and mechanical injury at collar region favour the disease incidence. Brownish lesions first occur on collar regions, which spreads to the entire pseudostem and cause complete yellowing of the plant. In severe case, the plant collapses leading to complete crop loss.

**Management :** Use disease free planting material, remove infected plant materials, improve drainage conditions, incorporate organic amendments like neem cake, drench the soil with Carbenilazim or apply biocontrol agents like *Trichoderma harzianumI* @ 2.5 kg/ha mixed with 50kg of FYM (lg/l of water).

**Field Inspection: :** A minimum of three inspections shall be made, the first after about 90 days, the second after about 150 days and the third after about 200 days of planting or at appropriate growth stage depending on the crop duration of the variety concerned to verify off types and other relevant factors.

| Factors | Foundation seed | Certified seed |
|---|---|---|
| Isolation distance | 5 m | 5 m |
| Off-types | 0.050% | 0.10% |
| Plants infested with scale insects | None | None |

**Harvest :**Harvesting is done on 8 months after planting and particularly during January - February months. Drying of stem and leaves indicates the harvesting stage in elephant yam.

**Yield**: The crop can yield about 30–35 t/ha in 240 days.

For seed purpose, the yams can be left in the field itself till planting the next crop or the lifted yams can be stored in sand or paddy straw.

## Seed Standards for Foundation and Certified classes

1. Seed size (weight of the tuber): 100-150 gm.
2. In a seed lot, tubers not conforming to specific size shall not exceed more than 5.0% (by number).
3. The seed material shall be reasonably clean, healthy and shall conform to the characteristics of the variety. The tubers not conforming to the varietal characteristics shall not exceed 0.050% and 0.10% (by number) for foundation and certified seed classes respectively.
4. Cut, bruised, unshapy, cracked tubers or tubers damaged by insects (other than scale insects) slugs or worms shall not exceed more than 1.0% (by weight).
5. Maximum tolerance limit of tubers showing visible symptoms of infestation caused by scale insects will none.

# 18

# Cabbage (*Brassica oleracea var. capitata* Linnaeus Cruciferae)

**Botany :**The cabbage plant, *Brassica oleracea*, is an herbaceous annual or biennial vegetable in the family Cruciferae grown for its edible head. The head of the cabbage is round and forms on a short thick stem. The leaves are thick and alternating with wavy or lobed edges and the roots are are fibrous and shallow. The plant produces large yellow flowers. The densely leaved heads can range in size from 0.5 to 3.6 kg (1-8 lb) depending on variety. The plant is usually grown as an annual.

## Climatic Requirements

Cabbage thrives in a relatively cool, moist climate with moderate rainfall, well distributed during the growing season. It can withstand frost in the head stage. It requires a dormant period of cool temperature to bolt and initiate seed stalks and flower. Cool temperatures, however, are effective only after stem diameter, is one cm at least. In temperate climates, this occurs during the winter after the first season growth. Flowering and seed production follow in the second year. Headed plants form seed stalks when exposed to mean temperature of about 50 °C for six to eight weeks. In India, seed production of cabbage is possible only in hilly areas.

## Land Requirements

Land to be used for seed production of cabbage should be free from volunteer plants.

## Methods of Seed Production

Being a biennial, cabbage requires two seasons to produce seed. In the first season, the heads are produced and in the following seasons seed production follows. The seed crop can be left in situ or transplanted during autumn. In situ method is usually followed for certified seed production and the later for nucleus seed production. *In situ* method, the crop is allowed to over winter and produce seed in their original position, *i.e.* where they are first planted in the seedling stage. In the transplanting method, the mature plants are uprooted. After removing whorls, the plants are immediately reset in a well prepared new fields in such a way that he whole stem below the head goes underground with the head resting just above their surface. There are three methods to produce seed of cabbage.

### Stump Method

In this method, when the crop in the first season is fully mature, the heads are examined for true to type. The plants with off type heads are removed. Then heads are cut just below the base by means of a sharp knife, keeping the stem with outer whorl of leaves intact. The beheaded portion of the plant is called "stump". The heads are marketed and the stumps either are leaf in situ replanted in the second season i.e. during autumn. The following spring, after the dormancy is broken, the bud sprout forms the axils of all the leaves and leaf scars.

#### Advantages

(i) Gives extra income by way of sale of heads.

(ii) The crop matures 12-15 days earlier than the head intact method and

(iii) Seed yield is slightly increased.

#### Disadvantages

In this method, flowering shoots are decumbent and require very heavy staking otherwise they breakdown very easily while interculturing or spraying.

### Stump with Central Core Method

In this method, when the crop is fully mature in the first season, the heads are examined for true to type. Plants with off type heads are removed and rejected. Then the heads are chopped on all sides with downward perpendicular cuts in such a way that the central core is not damaged. This is an improvement over stump method in that the shoots arising from the main system are not decumbent. During the last week of February and until 15th March, when the heads start bursting, two vertical cross cuts are given to the head. Taking care that the central growing point is not injured. In the absence of such cuts, the heads burst out irregularly and sometimes the growing tip is broken. The operation is completed by going around the field twice or thrice during this period.

### Advantages

(i) Shoots arising from the main stem are not decumbent, hence very heavy staking is not required and

(ii) Seed yield is increased.

### Disadvantages

The chopped heads cannot be marketed.

### Head Intact Method

In this method, when the crop is fully mature in the first season, the heads are examined for true to type. The plants with off type heads are removed from the field. The head is kept intact and only a cross cut is given to facilitate the emergence of a stalk.

### Advantages

(i) The removal for heads (stump method) or chopping of heads on all sides (central core intact method) is not required. This saves time and labour.

(ii) Very heavy staking is not required.

### Disadvantages

The seed yield is slightly low as compared to stump, or stump with central core intact method.

## Brief Cultural Practices (*in situ* Method)

### Time of Sowing and Transplanting

The sowing time of different varieties should be so adjusted as to complete head formation by the end of October or first week of November, at that time, the mean temperature falls to 10o C or below, at this temperature, the heads stand best for over wintering. Early varieties like 'Golden Acre' should be sown from 10th to 25th July and transplanted when the seedlings are three to four week old, during the second fortnight of August. This sowing time must be strictly adhered to, as the crop from the early sowings has matured head during September (20°C). The heads get infected with bacterial stock rot, which sometimes is very severe. The late crop, planted during September does not form heads and bolts directly during spring and the seed grower is not able to ascertain purity of the crop. Medium late varieties like Bruppe's Sure Head, and alter varieties like Drum Head, which takes about 2 to 3 months to produce mature heads, should be sown during the second and first fortnight of June, respectively and transplanting finished by the first week of August. The mean temperature 22.5°C, 20°C and 14°C of August, September and October, respectively, afford optimum requirements for growth and head formation. The later transplanted crop starts head formation during spring and continues up to June and usually does not produce seed stalks.

### Method of Nursery Sowing

The seeds are sown in raised nursery beds in a manner as described earlier in commercial production of cabbage.

### Source of Seed and Seed Rate

Obtain nucleus/breeder's/foundation seed from source approved by a seed certification agency. For main season and alter varieties, 375 to 400 g seeds/ha and early varieties, 600-700 g seeds/ha.

### Land Preparation

Prepare the land to a fine tilth by repeated ploughing and harrowing followed by leveling.

### Manure and Fertilizers

Cabbage grows satisfactorily only when the supply of organic matter is liberal. For good crop, apply 500 to 600 quintal of farmyard manure per hectare at the time of land preparation. Apply 100 kg/ha nitrogen, 60 kg/ha phosphorus and 60 kg/ha of potash by drilling, or by broadcasting, sufficiently before transplanting the seedlings. Give another dose of 50 kg/ha nitrogen as surface application at the time of seed stalk emergence during March. Extra application of nitrogen may be given as and when there is a need before flowering starts, depending upon the condition of the crop.

### Transplanting

Three to four week old seedlings are transplanted. Transplanting should preferably be done in the evening and the field irrigated immediately afterwards.

### Spacing

Late varieties - 60 x 60 cm; Medium varieties 60 x 45 cm; Early varieties 45 x 45 cm.

### Irrigation

Cabbage requires a continuous supply of moisture. Irrigate the crop as frequently as required. Heavy irrigation should, however, be avoided when the heads have formed. A sudden heavy irrigation after a dry spell may cause bursting of heads. Hoeing and Weeding At least three weddings and hoeing till the end of October are essential. One weeding and earthling up during November and December and the second during March when seed stalks have emerged, control weeds and also help in proper drainage during winter and thereafter. Staking After the flower stalk are sufficiently developed, staking is necessary to keep the plants in an upright position.

## Handling the mature head

After the planted crop has fully developed heads at the close of autumn, the next step is the handling of these plants for seed production. Handling of plants can be done by any one of the three methods i.e. stump, stump with central core intact methods, described earlier.

## Rouging

The first rouging is done at the time of handling of the mature heads. All off type plants, diseased, or otherwise undesirable types, are removed at this stage. The second rouging is done before the heads start bursting. The loose leafed, poorly heading plants and those having a long stem with heavy frame must be rouged out at his stage. It is highly undesirable to keep such poor plants in the seed plots. Subsequent rouging for off types, diseased plants affected by, black leg, soft rot or leaf spot should be done from time to time as required.

## Seed Certification Standards

Field Inspection: A minimum of three inspections should be done, the first before the marketable stage, the second at the marketable stage and the third at flowering stage.

## Field standards

| Factor | Foundation seed | Certified seed |
|---|---|---|
| Isolation distance | 1600 m | 1000 m |
| *Off-types | 0.10 | 0.20 |
| **Plants affected by seed borne diseases | 0.10 | 0.50 |

* Standards for off type should be met at and after flowering and for seed borne diseases at final inspection.

** Seed borne diseases shall be Black leg (*Leptosphaeria maculans* (Desm.) Cos. & de Not). Black rot (*Xanthomonas campestris* cv. *campestris* (Pamm.) Dawson), Soft Rot (*Erwinia carotovora* L.R. Jones).

## Harvesting and Threshing

Cabbage starts seed stalk elongation from 10-20th March when the mean temperature rises to 10-13°C. Flowering and pod formation starts during the first week of April at mean temperature of 13-18.5°C. From 15th April to 15th May, the crop is in full flush of flowering and fruiting. The ripening of pods commences by 15th June to 20th June and the harvesting continues up to second week of July. At mean temperatures below 20°C during June and July, the maturity of crop is delayed at least by a fortnight and the harvesting may continue up to July end. To avoid shattering of seeds, the whole crop is harvested in two or three lots with sickles. Generally, the early plants are harvested first, and when the pod colour is about 60-70 per cent of the rest of the crop changes to yellowish brown, it is harvested completely and piled up for curing. After 4-5 days, it is then threshed with sticks and sifted with hand sifters. After thoroughly drying, seeds are cleaned and stored.

**Seed yield:** 500-600 kg per hectare.

## Seed standards

The percentage of minimum physical purity of foundation and certified seeds should be 98% with a minimum of 65% of germination capacity and 7% of moisture content. The presence of inert matter should not exceed 2.0%.

# 19

# Cauliflower (*Brassica oleracea* var. *botrytis* Linnaeus Cruciferae)

**Botany**

Cauliflower is highly cross pollinated crop due to self incompatibility. Flower is protogynous in nature. Stigma remains receptive 5 days before and 4 days after opening of the flower. The time taken from pollination to fertilization is 248 hours depending upon the temperature. The optimum temperature for fertilization and seed development is 12°C – 18°C. Bees are the major pollinators.

**Method of seed production:** There are two methods of seed production

1. *in situ* method (seed to seed method)
2. Transplanting method (Head to seed method)

For seed production, seed to seed method is recommended since the head to seed method in India has not been very successful. In seed to seed method (In situ method) the crop is allowed to over winter and produce seed in the original position, where they are first planted in the seedling stage.

**Stages of seed production**

Breeder seed → Foundation seed → Certified seed

**Varieties**

**Early:** Early Kunwari, Pusa Katki, Early Patna, Pusa Deepali, Pusa Early Synthetic, Pant Gobhi3, Improved Japanese.

**Mid season:** Pant Shubhra, Pusa Synthetic, Pusa Shubhra, Pusa Aghani, Selection 235S, Hisar No.1, Pusa Himjyoti.

**Late:** Snowball 16, Pusa Snow ball 1, Pusa Snowball 2, PSK 1, Pusa hybrid -2

**Hybrids:** Pusa synthetic, Pusa hybrid 1 & 2

### Season

In the hills, the last week of August is the optimum sowing time. The seed is sown in a nursery and transplanting should be completed by the end of September. For early varieties (in plains) best season for sowing is the last week of May and transplanting should be completed during 152 week of July. In hills, sowing should be adjusted that the plants put up the maximum leafy growth by 15th December when the temperature goes down and plants become dormant for which last week of August is optimum and transplanting should be completed by the end of September. The mean temperature of 6.5 to 11° C during February to March is very conducive to curd formation.

### Land requirement

In the hills, select field on which the same kind of crop or any other cole crop was not grown in the previous two years, unless the crop within the previous two years, was field inspected by the certification agency and found not to contain seed born diseases infection beyond the maximum permissible limit.

### Isolation requirement

Cauliflower is mainly a cross pollinated crop. Pollination is chiefly done by bees. The seed field must be separated from fields of other varieties at least by 1600 M for foundation class and 1000 M for certified class seed production.

**Seed rate :** 375 to 400 grams/ha.

### Nursery

Seeds may be sown on raised nursery beds 15 20 cm height in rows with 10 cm spacing. Twenty five nursery beds of 2 to 2.65 M x 1 to 1.25 M size are enough for one hectare. Thin sowing should be done to avoid damping off. 3 T of FYM should be applied per 25 cents of nursery bed. DAP spray at 10 to 15 days after germination is important. Apply lime @ 5 t/ha before one month to nursery field and apply Borax and Sodium molybdate @ 4 kg/ha before sowing.

### Transplanting

Transplant the seedlings at 35 40 days old preferably at evening time with the spacing of 60 x 45 cm (for early varieties in plains) or 90 x 60 cm for late variety and irrigate immediately after transplanting.

### Main field manuring

The field should be prepared to fine tilt by deep ploughing and three to four harrowing followed by leveling. Cauliflower crop requires heavy manuring. Apply 50 60 tons of FYM/ha at the time of land preparation.

## Roguing

Minimum of four inspection are required *viz*., pre marketable stage, initiation of curd stage, curd formed stage and flowering stage. Roguing should be done based on the curd size, shape and colour, when fully developed. Off type plants with poor curd formation and plants affect by designated diseases like black leg, black rot, soft rot, leaf spot and Phyllody should be removed during roguing.

First roguing is done after curd formation. Plants forming loose ricey, fuzzy and buttons are rejected. Blind, deformed and diseased plants are also rejected. Second roguing is done after bolting but before flowering, plants with peripheral and uniform bolting are kept for seed production. Early and late bolters are also rejected.

## Field standards

| Factor | Foundation seed | Certified seed |
|---|---|---|
| Isolation distance | 1600 m | 1000 m |
| Off-types | 0.10 | 0.20 |
| Plants affected by seed borne diseases | 0.10 | 0.50 |

**Designated diseases:** Black leg, Black rot and Soft rot

## Pest and disease management

Use of insecticides during flowering affects the insect pollinators and will result in poor seed set. A single soil application of granulated Phorate, Dimethoate @ 18 Kg/ha during early February for control of sucking pests (Aphids) is advisable. In cauliflower the major disease in "Damping – off". Thin showing and Drenching with 150 g of Bavistin in 100 litres of water will control the disease.

## Scooping

Scooping the central portion of curd when it is fully formed helps in the early emergence of flower stalks in hills. Scooping is normally not required for seed production in plains.

## Curd pruning

The outermost curdles were pruned 5cms, away from the center.

## Half- curd removal

The curd was cut vertically into two parts from the middle and one of them was removed leaving half portion of the curd intact for seed production.

## Harvesting and Threshing

Harvesting is done when pods brown. Too ripe pods dehisce. Seed should not crush or split when rubbed between the hands. The harvesting maybe done in two lots. Generally the early plants are harvested first, when about 60 to 70 percent of the pods turns brown and the rest of the crop changes to yellowish-brown. After harvesting it is piled up for curing. After four to five days it is turned upside down and allowed to cure for another four to five days in the same way. It is then threshed with sticks and sifted with hand sifters. After thorough drying of seed in the sun (seven percent moisture content) it is cleaned and stored.

**Seed Yield:** Average seed yield varies from 250 to 400 kg per hectare

## Seed standards

The percentage of minimum physical purity of foundation and certified seeds should be 98% with a minimum of 65% of germination capacity and 7% of moisture content. The presence of inert matter should not exceed 2.0%.

# 20

# Carrot (*Daucus carota* Satova de Condle Umbelliferae)

Carrot (*Daucus carota* Stova condolle) is one of the most important and widely used root vegetable belonging to the family Umbelliferae. The seed production can be done during September – October in plains and in hills the sowing takes place in June and roots are replanted during the first week of October.

## Method of seed production

Carrot is a cross-pollinated crop and self-pollination occurs to the extent of 0-5%. Crosspollination is mainly through insects. Seeds should be allowed to set by cross-pollination in isolation. Seed production is done by seed to seed and root to seed method. In seed to seed method, the matured roots are left to produce flowers and seeds in the place where seeds are sown initially. In root to seed method, roots at edible maturity should be uprooted and the roots of true to varietal characteristics should be selected and transplanted to the well prepared field after proper trimming of roots and shoots. This root to seed method is preferred for seed production in carrot since the root rot infection is high in seed to seed method of seed production.

The isolation distance maintained between the fields of other varieties and the fields of the same variety not conforming to varietal purity requirements for certification is 1000 metres for foundation and 800 metres for certified seed production. During mother root production an isolation of 5 metres should be followed.

## Seed production stages

Breeder seed → Foundation seed → Certified seed

## Land selection

The land selected should be free from volunteer plants. The soil should be fertile and soft with good drainage facility. Since carrot is a cross-pollinated crop the land should be with pronounced isolation distance.

### Seed selection and treatment

Certified seeds should be obtained from an authorised source. Seeds should be healthy and free from disease and pest infection. Remove the broken, coloured seeds and use uniformly graded seeds. Seed rate is 1.5 kg/acre (4 kg/ha).

The selected seeds should be soaked in water for 72 hours and water should be changed every 24 hours. This method will remove all the germination inhibitors and improve germination. Seeds should be soaked in a solution of cow's urine (1 part cow's urine + 5 parts of water) for 30 minutes prior to the sowing. This will inhibit the seed borne diseases. Treat the seeds with *Trichoderma viride* @ 4 gms/kg of seeds.

The treated seeds are sown directly in the field ploughed for 3 – 4 times and formed into ridges or beds of convenient size. Well prepared soil of soft and smooth texture will enhance the germination and growth of the plant. Seeds are mixed with fine sand to facilitate uniform distribution and sown in ridges at 1.5 – 2.5 cm depth. After thinning the intra row spacing should be 5 – 10 cm. In replanting method, the mother roots are pulled out carefully without damage to the roots and selected based on the typical characteristics. Before replanting the shoot and the root parts are trimmed to 2/3 and ½ to ¾, respectively. The roots (also known as stecklings – roots used for replanting for seed production) are planted at a required spacing of 60 x 30 cm.

### Nutrient management

Farm yard manure or compost is applied @ 10 tonnes/acre (25 tonnes/ha) before last ploughing and incorporated into the soil. Neem cake @ 30 kg/acre (75 kg/ha) and vermicompost @ 250 kg/acre (600 kg/ha) should be applied as basal manure. Enriched vermicompost (2 kg *Azospirillum*, 2 kg Phosphobacterium and 2 litres Panchagavya mixed with 250 kg vermicompost and kept covered for a week and then used) @ 250 kg/acre (600 kg/ha) should be applied 20 – 25 days after sowing as first top dressing. Second top dressing should be done 40 – 45 days after sowing using neem cake 15 kg and vermicompost 250 kg mixed with 200 gms of asafoetida per acre (35 kg neem cake + 600 kg vermicompost mixed with 500 gms of Asafoetida Ferula assafoeticla F. per hectare). During flower initiation stage 10% tender coconut solution (1 litre tender coconut water + 9 litres of water) should be sprayed.

### Weed management

Weeding at regular intervals is very important for the seed crop. The first weeding can be done 15 – 20 days after seed sowing / replanting. Periodical removal of objectionable weeds should be done.

## Irrigation

Regular irrigation is a must to maintain a moisture content of 60 – 80%. First irrigation is done at the time of sowing / replanting. Subsequent irrigation should be done at an interval of 10 - 15 days. In replanted method, first irrigation is done soon after the replanting followed by second irrigation 4 - 5 days after planting. Frequency of the irrigation depends on the moisture content of the soil. Irrigation should be stopped when lower few pods start drying. Irrigation should be done 3 - 4 days before uprooting.

## Pest and disease management

Diseases like *Alternaria* blight and powdery mildew are commonly affecting the crop.Treat seeds with hot water, use disease free healthy seeds.

## Roguing

Roguing should be done in all growth stages like vegetative stage, flowering stage, stock formation stage and pod formation stage. All the off-types, diseased plants, plants with thin roots, plants coming to early flowering etc., should be rogued off. The maximum percentage of off-types and roots not confirming to varietal characteristics permitted at the final inspection is 0.10% for foundation seed production and 0.20% for certified seed production.

## Field inspection

In carrot a minimum of six field inspections should be done during the mother root production stage and seed production stage. In mother root production stage, two inspections should be done. The first inspection at 20 – 30 days after sowing to check isolation, off-types and other factors and second inspection at the time of uprooting the roots to determine the true characteristics of the roots. In seed production stage four inspections are scheduled during the pre flowering stage followed by two inspections at flowering stage and fourth one during maturity stage to check isolation, off-types, designated diseases, varietal characteristics and other relevant factors.

## Field standards

| | Foundation seed | Certified seed |
|---|---|---|
| Isolation distance | 1000 m | 800 m |
| Off-types | 0.10% | 0.20% |
| Roots not confirming to varietal characteristics | 0.10% | 0.20% |

## Harvesting

Harvesting is done once the seed heads are physiologically mature. The physiologically mature seed heads will turn from green to brown colour. The matured seed heads should be harvested at once and dried further. Seed extraction and processing. The dried pods are crushed to separate the seeds, since the pods do not shatter. The separated seeds should be cleaned using BSS 12 wire mesh sieve.

## Drying and storage

The extracted seeds should be dried under the shade for a week or two to attain a moisture level of 8% before storage. Seeds can be stored in cloth bags or 700 gauge polythene bags. Under suitable storage conditions the seeds can be stored for about four years.

## Seed standards

The percentage of minimum physical purity of foundation and certified seeds should be 95% with a minimum of 60% of germination capacity and 8% of moisture content. The presence of inert matter should not exceed 5.0%.

# 21

# Beetroot (*Beta vulgaris* Linnaeus Chenopodiaceae)

## Origin

Beet root originated from *Beta vulgaris* L. ssp. maritima by hybridization with B. patula. Crop has site of origin probably in Europe. Earlier types were with long roots like that of carrot. Beet root, sugar beet and palak belong to species *B. vulgaris* and are cross compatible.

## Botany

Beet root is a biennial, producing a fleshy elongated hypocotyls and a rosette of leaves in first year and flowers in second year. Upper portion of fleshy root develops from hypocotyls and basal part from tap root. Concentric rings seen in cross section of root are as a result of alternate formation of vascular tissues and storage parenchyma tissues. Root skin colour varies from orange red to dark purple red. Colour of beet root is due to presence of red violet pigments of â cyanins and a yellow pigment, â xanthin. It has extensive tap root system and tap root grows as deep as 3m. Rosette leaves develop in a close sp

## Climate

Beet root is hardy to low temperature and prefers cool climate. Though it grows in warm weather, development of colour, texture, sugar content etc. of roots is the best under cool weather. High temperature causes zoning *i.e.*, appearance of alternate light and dark red concentric rings in the root. Extreme low temperature of 4.5-10.0 °C for 15 days will results in bolting. It requires abundant sunshine for development of storage roots.

## Soil

Deep well drained loam or sandy loams is the best for beet root cultivation. Heavy clayey soils result in poor germination and stand of crop due to formation of a soil crust after rains or irrigation. Roots may be misshaped and will not develop properly in heavy soils. Beet root is highly sensitive to soil acidity and the ideal pH is 6-7. Beet root is one of a few vegetables which can be successfully grown in saline soils.

### Varieties

Beet or Garden beet (*Beta vulgaris* L.) Beet has only european type of varieties. These are Detroit Dark Red, Crimson Globe, Crosby Egyptian, Early Wonders

### Seed production

As with other biennial root crops, both seed-to-seed and root-to-seed methods can be followed in producing beet seed. Since the interior flesh colour is important, the root-to-seed method is followed greatly. The seed-to-seed method has certain limitations for its use as described under other root crops.

### Seed-to-seed method

This method is also known also *in-situ* or over-wintering method. It is much simpler than root-to-seed method because the roots are allowed to over-winter in the field where they were grown earlier and thus time and cost of labour is saved. The only drawback is the lack of thorough roguing. The sock seed used for seed-to-seed plantings should be taken from a crop which represents trueness-to-type and freedom from rogues of all kinds. The use of this method should be restricted to the production of only certified seed (market seed) as on the grower's field negligible off-types do not matter much.

### Root-to-seed method

This is the specialized technique of quality seed production where the roots / stecklings of beet are grown first year in the same way as for market crop and in second year the selected roots are replanted in the well prepared field for the seed crop. This is the most satisfactory method of stock seed production and also followed in breeders' seed and foundation seed production. When the roots reach full size by the end of the growing season, they are either replanted immediately or stored in trenches depending upon the severity of winter (amount of snowfall). In areas of heavy snowfall like Kalpa (Himachal Pradesh), stecklings uprooted in November-December are stored in shallow trenches up to March by the time snow melts. The stecklings receive the necessary chilling during their storage in the trenches. The stecklings should be larger in size as smaller ones get shriveled and dried in the storage. The depth of the trench should not be more than 45-60 cm otherwise the stecklings in the lower layer of storage pit get overheated and will be devoid of adequate thermal induction although they are well protected from freezing. In Kullu and Srinagar, where the snowfall is light the roots are planted immediately after lifting. The merits and demerits of this method remain the same as with other root crops.

## Land preparation

The soil should be thoroughly prepared by ploughing the land about 4-5 weeks before sowing. Ploughing in June in temperate region before the rainy season starts, helps to take care of weeds and ease the preparation of field in July / August when rains actually start.

## Manures and fertilizers

Apply well rotten farmyard manure 10-15 tonnes / ha atleast 4-6 weeks before sowing the seed. At the field preparation, 50 kg N, 50 kg $P_2O_5$ and 100 kg $K_2O$ should be incorporated thoroughly into the soil in the ratio of 1:1:2. Another dose of N 50 kg / ha is applied before initiation of root development. If the seed is to be produced by *in situ* method apply 10 tonnes farmyard manure, 50 kg $P_2O_5$ and 50 kg $K_2O$ at the field preparation. Apply about 100 kg N in 2 equal split doses, first after 40 days when the new leaves start developing and the second at bolting of the crop. In the crop raised for steckling production, borax @ 20-25 kg/ha is helpful to prevent canker and internal breakdown as black spot or dry rot and the development of good quality roots. In poor growth, 4-5 weekly sprays of 1-2% urea in the replanted crops are helpful to push up the growth.

## Spacing

For beet seed production, moderate-sized stecklings having a diameter of 5-7 cm are more preferable to either smaller or larger ones. For raising of stecklings, 45 cm x 10 cm and for replanting stecklings 45 cm x 45 cm of 60 cm x 45 cm spacings are most suited.

## Seed rate

For *in-situ* method of seed production, seed 5-6 kg/ha is sufficient. For raising stecklings, seed 10-12 kg/ha is used and this will provide sufficient stecklings under better management to replant 7-8 ha of seed crop.

## Sowing time and replanting

The optimum time for sowing is the second fortnight to July in Kullu and Srinagar where the snowfall is comparatively low and root-to-seed or *in situ* method is followed. Time of sowing influences size, shape and colour of foliage, and texture and become thick, broad and savoyed whereas high temperature of 10°-15°C become thick, broad and savoyed whereas high temperature of 21.1° – 26.7°C leads to the development of narrow, thin and smooth - surfaced leaves. At high temperature, the roots become coarse with woody flesh and dull colour. Hence, early sowings should be discouraged for seed crop. Transplanting of mature best stecklings like other root crops, should be done in November or early December after which the temperature drops considerably. Late planting results in the development of small seed-stalks with less number of branches and flowers.

## Bolting, flowering and seed-setting

Beet has only temperate types which bolt and produce seed only in response to low temperature. Sometimes premature bolting is there in the crop grown for stecklings. There can be two causes responsible for it: (a) seed produced in areas having long winters and short summer with good rainfall get vernalized during later stages of ripening. The stecklings raised from this seed go to bolting before attaining maturity : (b) if the temperature goes low (4.4° – 10°C) for 15 days or more in the early stages of crop growth, some percentage of bolting is likely to occur in the field. This is particularly true in the case of the crop grown later in the season (October-November) in the plains of North India, which is vernalized during winter before reaching maturity.

Transformation from vegetative to reproductive phase takes place only when the plants after root formation are exposed to low temperature of 4.4°–10°C for 60-90 days, depending on the variety. Crop is allowed to over winter either in the field or in the trenches depending on the extent of snowfall. At lower temperatures induction period is reduced to some extent. Unless the chilling requirement is complete, a rise in temperature beyond 21°C nullifies the effect of earlier cold period and causes reversion of reproductive processes to the vegetative phase. This is called devernalization. As this reversion is irreversible, such plants cannot be vernalized again and remain in the vegetative stage. Plants are conditioned for production of flower-stalks by low temperature treatment but the elongation of seed-stalks actually takes place under long photoperiods.

## Field standards

| Factors | Foundation | Certified |
|---|---|---|
| Isolation distance | 1600 m | 1000 m |
| *Roots of other varieties not conforming to varietal characteristics | 0.1% | 0.2% |
| Off types | 0.1% | 0.2% |

## Harvesting

Ripened seeds sometimes shatter easily but if harvesting is done early in the morning when it is sufficiently moist the problem of shattering can be prevented. As ripening is uneven, 2 to 3 harvestings manually are often necessary. To prevent losses in shattering, the seed crop should be harvested when two-thirds of the seed on a branch are being ripened i.e. changing in colour to light brown.

## Seed yield

On an average 800 – 1,000 kg/ ha seed yield can be obtained.

## Seed standards

The percentage of minimum physical purity of foundation and certified seeds should be 96% with a minimum of 60% of germination capacity and 10% of moisture content. The presence of inert matter should not exceed 4.0% and the seeds of other crop varieties should not be more than 5/kg of foundation seeds and 10/kg of certified seeds.

# 22

# Peas (*Pisum sativum* var. *arvense* Linnaeus *Papilionaceae*)

**Botany :** Pea is a self pollinated crop. Time of anthesis is 5.45 to 6.30 AM. Time of dehiscence is 5.45 to 7.30 AM. Duration of pollen futility is on the day of anthesis to 24 hr there after. Duration of stigma receptivity is 48 hr before and 24 hr after anthesis.

**Method of seed production :** Var: Open pollination ; Hybrids: Emasculation and Pollination

**Varieties / Hybrids**

**Early varieties :** Arlcel, Matar Ageta 8, Early Badger, Pant Uphar, Hisar Harit, P.M. 2, Jawahar Matar 4, Meteor, Alaska, Lucknow Boniya, Pant Sabzi Matar 3, Pant Matar 2, VL Ageti Matar 7, Asauj.

**Mid season varieties:** Bonneville, Lincoln, VL 3, Shalimar Matar, Solan Nirog, Jawahar Matar 1, Pant Uphar, p88, Khapar Kheda, JP. 83, JP71, NP. 289, T19

**Late varieties:** Kinnauri

**Isolation requirement :** Minimum isolation distance of 10 m for foundation seed class and 5 m for certified seed class from other pea field is necessary for certified seed production.

**Season:** Middle October to November in the plains

October to November, February to March in the hills

**Field preparation :** The field should be thoroughly ploughed for fine filth and apply 25t of FYM/ha. At the time of sowing 50kg P and 50kg K should be applied. Sowing should be done in rows with seed drill or behind the plough at 3 45 cm depth with the sparing of row to row 45 to 60 cm and plant to plant 5 cm.

**Top dressing:** Top driving with 25 kg of N/ha in two doses driving early growth period and flowering time should be given.

**Irrigation :** 2 3 irrigation at an interval of 15 to 20 days should given. Irrigation at the time of flowering and fruiting is must.

**Interculture:** One hand weeding after 3 4 weeks of sowing should be done

**Pest and disease control:** The important pest attacking in peas are aphid, pod borer and it can be controlled by spraying Cypermethrin or permethrin or Imidacloprid systemic insecticide. Spraying of Karthane (0.2%) should be done for control of Powdery mildew disease.

**Roguing:** A minimum of three inspections are to be done, first before flowering, second at flowering and third at edible pod stage. The off type plants and diseased plants affected by pea mosaic, foot rot and blight should be rogued out from the seed field.

**Harvesting :** The harvesting should be started when 90% of the pods turn brown. The plants should be uprooted, stocked in small heaps and allowed to dry in the field for a week.

**Threshing :** Threshing should be done by beating the pods with sticks or thresher is used, its cylinder speed should be reduced to minimize the mechanical damage. After threshing, severely damaged seeds, split seeds and shrunken seeds must be removed. Sieve size is 9.5mm round holed sieves.

**Seed yield:** 20 25 H/ha

**Seed standards:** The percentage of minimum physical purity of foundation and certified seeds should be 98% with a minimum of 75% of germination capacity and 9% of moisture content. The presence of inert matter should not exceed 2.0%.

# 23

# French bean (*Phaseolus vulgaris* Linnaeus *Papilionaceae*)

**Botany:** French bean has tap root system with poor nodule formation. Leaves are trifoliate. Though a self-pollinated crop, French bean offers wide variability for plant growth (bushy / climbing), colour of pod (green / waxy coloured), cross section of pod (flat / oval / round), pliability (stringed / string less) etc.

**Varieties**

1. Arkakomal: Plant is bushy. 70 days for flowering.
2. Blue pod medium: It is a white seed variety. It has bluish black spots on pods.
3. Bountiful: Introduction from US comes up in September to February. Seeds are Brown sweetish. It is resistant to common disease.
4. Contender: Tolerant to powdery mildew and mosaic.
5. Pant anupama: Developed from pantnagar. Bushytype
6. Pusaparvathi: Developed by X ray radiation of American variety IARI

**Soil:** French bean can be grown on wild range of soils, sandy loams are best. Heavy soil gives good yield. Ideal pH range is between 6 to 7. French bean mature early on sandy soils compared to other soils.

**Climate:** French bean is a cool season crop. It gave good yield under mild warm season. It is sensitive to frost. French bean has the cultivars of long day, short day and day neutral varieties. Most of them are day neutral. Seeds will not germinate very high temperature. Continuous rains, resulting in the breakage of pods.

**Season:** In plains of North India, French bean is sown during two seasons *viz*., July-September and January-February. In hills, sowing is done from March to May.

**Land preparation and sowing**:Land is ploughed to a fine tilth and divided into plots of convenient size. Ridges and furrows are prepared by ploughing after a basal dose application of farmyard manure. Field is irrigated once and seeds are sown under optimum moisture condition on side of ridges 2-3 days after irrigation. Spacing and seed rate vary with varieties. Early varieties are sown at a spacing of 45-60 cm x 10-15 cm and seed rate required is 80-90 kg / ha. Pole types are sown at 1.0 m apart in hills @ 3-4 plants / hill and seed rate is much less (25-30 kg/ha.).

### Seed rate

Cultivars spacing

Bush type of cultivars - 95 kg/ha

Pole type of cultivars - 25 to 30 kg/ha

### Spacing

Bush type - 60 x 15 cm

Pole type - 1 m x 60 – 75 cm x 30 cm Seeds are sown 5 cm deep.

Two seeds are sown at one place and one of the weak seedlings is removed after completion of germination. Seeds germinate slowly, at soil temperature of 15°C and they germinate 10 days after sowing under favourable conditions.

**Irrigation:** French bean is said to be a drought resistant crop. For better growth and yield proper supply of moisture is essential. Irrigation is required in the early phases of crop growth during blooming and pod development period. Plants are susceptible to water stress. Irrigation at regular intervals are necessary. Lack of adequate soil moisture results in reduced percentage of pod setting, reduced length of pods, reduced number of seeds per pod and high fibre content in pods.

**Manure and fertilizers:** Apply FYM 25 t/ha at the last ploughing. N 90 kg and P 125 kg/ha should be applied on one side of the ridges. For rainfed conditions of Shevaroy hills, apply as a basal dose of 62.5 kg/ha of Phosphorous as super phosphate and with another half of 62.5 kg/ha Phosphorous as FYM enriched super phosphate.

**Intercultural operations:** French bean is a shallow rooted crop and only light inter-cultural operations are practiced. During early stages of crop, weeding followed by fertilizer application and earthing up can be synchronized. A pre-sowing application of Fluchloralin @ 2.1 /ha checks weed growth for 20-25 days. Water stress influences yield of French bean and crop is most sensitive at flowering and fruiting stages.

6-7 irrigations are required during growing season. Staking is an important operation for pole types and bamboo sticks or any locally available materials should be erected when plants start vining. Individual vertical stakes and horizontal canes at 40 cm distance are erected for encouraging growth and spread of plants. Application of plant growth regulators like PCPA (2 ppm) and NAA (5-25 ppm) has favourable effect on fruit set and yield.

**Harvesting and yield:** Pods are harvested at full grown stage but immature and tender. Pods are ready for harvest 7-12 days after flowering depending on varieties. In bush varieties, 2-3 harvests and in pole types 3-5 harvests are made. Quality of beans varies with harvests and best quality fruits are obtained in initial harvests compared to later harvests. Loss of crispness during storage and in last harvest is attributed to loss of water and increase in water soluble pectin. Seed weight is a major indicator of green bean harvest maturity. Yield of tender pods varies from 8-10 t/ha in bush varieties and 12-15 t/ha in pole types. Dry beans are harvested when majority of pods are fully ripe and colour turns yellow. Seed yield varies from 1250 to 1500 kg / ha.

Pests and diseases Crop is affected by pests like stem fly, thrips, mites, bean beetle, bean weevil, aphids etc. Yellow mosaic, anthracnose, powdery mildew, rust, root rot and wilt and leaf spot are common diseases affecting French bean.

## Average yield

Bushy varieties is 4 to 5 t/ha

Pole varieties is 7 to 10 t/ha

Yield of dry seeds varies from 1.2 to 1.8 t/ha

## Storage

It is completely dried seeds are stored in glass container with tight podding varieties or polythene bags.

## Seed standards

The percentage of minimum physical purity of foundation and certified seeds should be 98% with a minimum of 75% of germination capacity and 9% of moisture content. The presence of inert matter should not exceed 2.0%.

# 24

# Potato (*Solanum tuberosum* Linnaeus Solanaceae)

Potato has special seed production problems as the quality characteristics of seed potatoes are influenced by a number of factors and important amongst them are the diseases and pests namely, viruses, fungi, bacteria and nematodes. Once the seed tuber is infected the pathogens, specially viruses, permeate through the plant system. The plant growth, therefore declines and the yield reduces progressively. It is therefore, important that seed stocks should not only be genetically pure but also should be in right physiological condition and disease free at the time of planting. Seed production technology developed for seed potato production, thus aims at producing disease free, genetically pure seed. There are now two independent channels of seed production for hills and plains.

**Hill seed :** The seed produced in hills (2500 metres above sea level) at suitable locations is called 'Hill Seed'.

**Plain seed :** The seed produced in plains at suitable locations is called 'plain seed'. Northern plains have emerged as an important source of potato seed production. The low aphid, plains seed is in right physiological condition at the planting time and yields higher than the traditional hill grown seed.

Once healthy seed potatoes are introduced into the system of growing them during low Aphid period accompanied by a systematic insecticide application, roguing and removal of haulms before the aphids attain critical number and the re growth is checked, the health standards for the seed crop could be maintained for a number of generations. This system of seed potato production has been designated as 'Seed Plot Technique'.

**Stages of seed production :** For seed multiplication and certification purposes following stages are recognized. BS – FS I – FS II – CS I – CSII.

**CS 11:** This is done in case of those varieties which have a low rate of multiplication and in years of shortage of seeds.

**Land requirements :** A crop of seed potato shall not be eligible for certification if grown on land infested with wart and/or cyst forming nematodes; or brown rot or non cyst forming nematodes within the previous three years; and common scab. Preference should be given to two to three year crop rotation.

**Isolation requirements :** A minimum isolation distance of 5 m for foundation and certified seed class should be provided all around a seed field to separate it from fields of other varieties, and fields of the same variety not conforming to varietal purity and health requirements for certification.

**Time of Sowing:** The sowing should be done from 20$^{th}$ September (when rainfall is low), or 25$^{th}$ September, up to 15$^{th}$ October. Delayed plantings will result in poor yields.

**Seed rate:** Seed rate depends upon tuber size. Twenty five to 30 qtl of seed potato per hectare will be sufficient if the usual sized tuber (4 to 6 cm) are used.

**Fertilization:** 125:80:100 kg NPK with 25 ton FYM/ha. Apply all phosphorus, potash and half of the nitrogen at the time of sowing. The remaining half of nitrogen should be applied about thirty five days after sowing, or when the plants are about to 30 cm high. For best results, the fertilizers should be placed either 5 cm below the tubers, or on the side.

Method of sowing**:** Whole tubers should be used for planting. Tubers should be under sprouting (sprouts 0.5 to 1 cm long) for quick emergence. After 15$^{th}$ October when the temperature goes down, cut tubers can also be used for planting. Care must be taken that each piece to be used for planting has two or three emerging eyes and weighs at least 40 gm. By this practice the seed rate is reduced considerably. Plant the tubers 3 to 4 cm deep in the soil having adequate moisture. Row to row spacing at 60 cm and tuber to tuber spacing at 15 to 20 cm is recommended.

**Irrigation:** Potato requires light and frequent irrigation. First irrigation should follow immediately after emergence. Subsequent irrigations should be given at proper intervals. Restrict the irrigation after the crop has tuberised well. Withhold irrigation by the third week of December i.e., ten to fifteen days before cutting of haulms.

**Interculture:** Keep the field free from weeds. At least one earthing up is a must. It should be done when plants attains the height of 15 cm.

**Haulm cutting:** The practice of haulm cutting is adopted as a precautionary measure to avoid chances of viral disease transmission through the vectors like aphids. The haulms must be cut by the end of December, or at the latest by the first week of January before the aphid population reaches the critical stage (20 aphids per hundred compound leaves). No re-growth should be allowed.

**Roguing:** Very careful roguing is required for producing a high quality crop of seed potato. The roguing is to be done at the following stages:

**First roguing:** First roguing should be done 25 days after sowing to remove:

a) All virus affected plants and b) All plants apparently belonging to other varieties and which can be identified from foliage.

**Second roguing:** It should be done when the crop is fully grown. This would be about 50 to 60 days after sowing. At this time tubers are formed and therefore, while roguing, not only the upper portion of plant, but all the tubers belonging to the plant should be removed carefully. Also at this stage the virus affected plant as well as off type, should be removed.

**Third roguing:** This is the third and final roguing and should be done just before cutting the foliage. Foliage should not be cut unless this roguing has been completed. At this stage, all virus affected plant and off type plants, along with their tubers have to be very carefully removed.

## Harvesting

a. The crop is ready for harvest ten to fifteen days after haulm cutting when the skin of tuber has hardened. Premature harvesting causes handling problems, as the soft skin gets easily peeled of and further such tubers cannot withstand long transportation and storage.

b. At the time of potato digging, the moisture in soil should be optimum for obtaining clean tubers.

c. The harvesting of seed potatoes can be done by any of the equipment available in the market for this purpose. Every effort should be made to avoid cuts, bruises, etc. After harvesting, tubers should no be left exposed to the not sun for a prolonged period (not more than an hour). It should be immediately lifted and carried to an airy shed and kept in piles (height 1 m, width 3 m) for 7 to 10 days so that the superficial moisture evaporates and further hardening of skin is achieved. If sheds are not available, piles may be made infield and covered with dry haulms.

**Sorting and Grading :** When the potatoes are properly cured, grading should be gone. A single grade from 3.0 to 5.5 cm is being made at present for 'Plain Seed' by hand grading. While grading, the shape; colour depth of eyes, etc. of tubers should be critically examined and of types discarded. In addition to off types, the tubers with cuts, bruises, cracks or otherwise mechanically damaged or showing visible symptoms of late blight, dry rot, charcoal rot, wet rot, scab, black scurf, etc. should invariably be removed.

**Seed standards:** Size of seed corms: 4 6cm x 2.5 to 3.5cm ; weight : 20 40g

**Packing:** After sorting and grading the seed potatoes should be put in clean jute hessian bags (50 kg size) and the bags appropriately labelled.

**Storage:** Soon after packing, the seed potatoes should be moved to the end use areas for cold storage. If the ambient temperatures are above 32°C, the seed potato should firstbe kept in pre-cooling chambers, or in a cool place for preconditioning, and then stored in cold storage at temperatures from 2.2 to 3.3° C and 75 to 80 % relative humidity. Periodic inspection of seed stocks in cold storage is necessary, to ensure that stocks are keeping good. Turning of bags during rainy season helps in improving aeration.

Potato is traditionally grown vegetatively through seed tubers. This results in continuous accumulation and increase of various tuber borne diseases in seed tubers and consequent reduction in crop yields. To overcome these problems a new potato production technology making use of True Potato Seed (TPS) as planting material for raising the crop has been developed. TPS can serve as a cheap and highly productive planting material for raising commercial potato crop, especially in areas where good quality seed tubers at reasonable prices and in adequate quantities are not available.

The major advantages of this technology over the traditional seed tuber technology are as follows:

1. Unlike the seed potato production which is confined to northern India only, the TPS can be produced in all potato growing regions.
2. The crop raised through TPS is almost disease free as most of the diseases get filtered out during TPS production.
3. The TPS being very small, can be stored and transported easily, whereas the seed tubers are bulky hence the storage and transport are expensive.
4. TPS provides a low cost potato production technology where only about 50 g TPS is required for sowing in about 375 m$^2$ area for producing seedling tubers enough for planting one hectare next year. About 150 g TPS/ha is required if the commercial crop is to be raised in the first year itself by transplanting seedlings in field.

The TPS technology involves two major steps *viz.,* (a) production of hybrid seeds, and (b) its used as planting material for raising the commercial crop. Two high yielding TPS hybrids *viz.,* TPS C 3 and HPS 1/13 have been recommended for commercial use.

## Hybrids and their parents:

TPS C 3 : JTH/C.107 X JEX/A 680 16
(Female x Male)

HPS 1/13 : MF.1 x TPS 13
(Female x Male)

## Production of Hybrid TPS

The hybrid TPS can be produced both in the hills, where the crop is grown during long summer days and also in the plains where the crop is grown in winters. Whereas, in the hills the crop flowers natural and hybridization for seed production can be done easily; in the plains, however, additional light has to be provided to induce flowering in the crop. In Northern plains, where the winters are severe, the parental lines need to be planted either about 15 20 days before optimum time of the planting of the crop, or the planting is delayed till the second week of November so that the flowering time does not coincide with the severe winter period. Delayed planting of parental lines should not be done in frost prone areas.

## Following steps are involved in the production hybrid True Potato Seed (TPS)

1. If the TPS parents are planted in the plains, there is generally need to provide extra light for about 5 hours at the end of the day to prolong the day length and get proper flowering. Arrangements can be made for providing light from 150 W Sodium vapour lamp (one for about every 100 sq.m). Light arrangement is not required if the parents are planted (in April/May) in the northern hills during summer crop season.
2. Plant male and female parental lines (TPS 13 as male and MF 1 as female for producing hybrid TIPS HIPS V13, and JEX/A 680 16 as male and JTH/C 107 as female for producing hybrid TPS of C 2) in two separate but adjacent blocks. The area required for planting male block is generally kept at about 1/4 to 1/6 of the female block, depending on pollen producing ability of the male parent.
3. Plant the male block about a week before planting female block during the main crop season in the plains. Follow the spacing of 60 x 20.cm.
4. In the female block, prepare beds of three rows each. For this, draw 3 rows at 50 cm inter row distance leaving 80 cm walking space between two adjacent beds. Plant tubers at 15 cm intra row distance.
5. Use about 30 g size seed tubers or the seed pieces for planting female block. After germination, trim the plants in female block to retain a single stem per plant.

6. Follow all other cultural and plant protection practices for ware potato crop.
7. In the plains, after the germination is complete, switch on the light in hybridization block in the evening before sunset for a period of 5 hours every day. This will facilitate rapid plant growth and flowering in the parental material.

## Hybridization

When the parental material comes to flowering, follow the steps given below to produce hybrid TPS

i. Trim the flower bunches in the female plants to retain only 6-8 large size buds per bunch. Very small buds, old flowers and berries if any, should be removed from bunches the prepared for pollination next day.
ii. Collect flowers from the male parent in the evening preceding the day of pollination. Only just opened flowers with anthers that are about to shed pollen or the large size buds, which would open next day, should be collected.
iii. Spread flowers of the male parent on a sheet of paper placed on the table a room temperature and use them for extracting pollen next morning.
iv. Extract the pollen in a small dish by shaking anthers using electric buzzer or manually.
v. Pollinate the flowers of the female parent by dipping the stigma in pollen or applying pollen to the stigma using a brush. Do the first pollination between 8 10 AM and re-pollinate the flowers at the same time next day. Continue the process of pollination till the flowering period of the crop is over. Usually there are two flushes of flowers during the crop period.
vi. Store the left over pollen in small vials, keep the vials in a dessicator and place the dessicator in refrigerator at 6-10° C for use next day, if necessary. It is advisable to use fresh pollen for pollination every day.
vii. Provide support (use sticks) to the stems of female plants.
viii. After berries are formed, the berry bunches should be covered with thin muslin cloth bags of about 8 x 12 cm size.

## Seed extraction and storage

i. Collect well developed berries about 50 55 days after pollination, keep them in trays and allow them to ripen at room temperature for a period of about 2 weeks.

ii. Mash the soft ripe berries.

iii. After mashing the berries, separate out the TPS with a high pressure water source. Treat the seed and pulp mass with 10% hydrochloric acid (HCI) with continuous stirring for 20 minutes. Wash the seeds with water at least 3-4 times to ensure complete removal of HCI.

iv. Immediately after washing, surface disinfect the seeds by soaking in 0.05% solution of Sodium hypochlorite for 10 minutes and again wash with clean water to ensure that there is no Sodium hypochlorite solution on seed.

v. Spread the seed in a thin layer on muslin cloth stretched over a wooden frame and keep the frames in a well ventilated room preferably under fan for 24 hrs, in shade. TPS can be safely dried in any type of low moisture environment *i.e.* forced air oven, fan etc. Temperatures above $30^0$C should be avoided during the initial seed drying period. Thereafter, seeds can be safely dried under temperatures not exceeding $40^0$C + $5^0$C.

vi. Expose the shade dried seeds to warm sun for about ½ hr to reduce the moisture content.

vii. Keep the seeds in a moisture proof container along with silica gel bags and store them at low temperatures. The seeds may be kept in polythene lined aluminum foil covers or double polythene bags or tin cans, sealed and stored in desiccators kept in refrigerator at 6 $10^0$C or even at room temperature during the winters.

# Section 2: Ornamentals

# 25

# Marigold (*Tagetes erecta* Linnaeus Compositae)

In India marigold is one of the most commonly grown flowers and used extensively on religious and social functions in different forms. Because of their ease in cultivation, wide adaptability to varying soil and climatic conditions, long duration of flowering and attractively coloured flowers of excellent keeping quality, the marigolds have become one of the most popular flowers in our country. Flowers are sold in the market as loose or as garlands. Due to its variable height and colour marigold is especially use for decoration and included in landscape plans.

### Varieties

There are 33 species of marigold and numerous varieties. There are two common types of marigold:

I) The African Marigold (*Tagetes erecta* Linnaeus)

II) The French Marigold (*Tagetes patula* Linnaeus).

**Origin:** African marigold: Mexico, French Marigold: Mexico and South America

Both have deeply cut foliage with a characteristic odour.

**I) The African Marigold** (*Tagetes erecta* Linnaeus)

The African Marigolds are generally tall (up to 90 cm) with large sized double globular flowers of lemon, yellow, golden yellow, primrose, orange or bright yellow colours. There are also dwarf varieties (20 to 30 cm) having large double flowers. The important varieties are: Giant Double African Orange, Giant Double African Yellow, Cracker Jack, Climax, Dubloon, Golden Age, Chrysanthemum Charm, Crown of Gold, Spun Gold.

**II) French Marigold** (*Tagetes patula* Linnaeus)

The French Marigolds are mostly dwarf, early- flowering and compact with dainty single or double blooms, borne freely and almost covering the entire plant. The colour flowers may be yellow, orange, golden yellow, primrose, mahogany, rusty red, tangerine or deep scarlet or a combination of these colours. The important varieties are: Red Borcade, Rusty Red, Butter Scotch, Valencia, Sussana. However, in the market mostly orange colour varieties are preferred and the variety which is dominating is African Giant Double Orange.

Local types (orange & yellow), Pusa Narangi Gainda, Pusa Basanthi Gainda (IARI varieties) and MDU 1 can be cultivated.

### Soil and climate

The marigolds are hardy and can be successfully grown in different types of soils and climate. Marigold can be successfully cultivated on a wide variety of soil. The French marigold grows best in light soil while the African marigold requires a rich, well-manured and moist soil. However, the soil is deep fertile friable having good water holding capacity well drained and near to neutral in reaction viz.pH 7.0-7.5 is most desirable. They can grow in almost all seasons except in very cold weather, as they are susceptible to frost. Marigolds require mild climate of luxuriant growth and profuse flowering. For seeds germination optimum temperature ranges 18 to $30^0$ C. Soil and planting is carried out during rainy season winter and summer season hence flowers of marigold can be had almost throughout the year.

### Seeds and sowing

The seeds are sown throughout the year. Nursery is raised with 1.5 kg seeds/ ha and the seedlings are transplanted after four weeks on one side of the ridge at 45 x 35 cm spacing. Treat the seeds with *Azospirillum* (200 g in 50 ml of rice gruel) before sowing.

### Irrigation

Irrigation is done once in a week or as and when necessary. Water stagnation should be avoided.

### Manuring

During last ploughing, incorporate 25t/ha of FYM. Apply 45:90:75 kg NPK/ha as basal and 45 kg N/ha as top dressing 45 days after planting.

### After cultivation

Weeding should be done as and when necessary. Irrigation should be given immediately after planting and light irrigation on third day after planting. Water stagnation should be avoided. Based on the soil moisture condition, irrigation should be done.

### Nipping/tipping

Thirty days after planting terminal portion should be tipped / removed to encourage the branching.

### Pests and Diseases

Thrips and Caterpillar : 0.1% Phosphamidon or Imidacloprid spray

Black spot, leaf spot : 0.2% Dithane M 45

### Crop duration

The crop duration is about 130-150 days.

### Harvest

Flowers are picked once in 3 days beginning from 60 days after planting.

### Yield

The average yield is about 18 t/ha.

### Seed standards

The percentage of minimum physical purity of foundation and certified seeds should be 97% with a minimum of 70% of germination capacity and 8% of moisture content. The presence of inert matter should not exceed 3.0%.

# 26

# Chrysanthemum (*Chrysanthemum segetum* Linnaeus Compositae)

**Botany :Annual *Chrysanthemum*** is a winter annual that can be grown either as pot plants or in groups as bed plants for various ornamental purposes. There are two types of annual *Chyrsanthemums*: *Chrysanthemum Segetum* and *C.coronarium*. Major varieties of *Chrysanthemum segetum* L. are Morning Star, Evening Star, Eastern Star, Blanca, Eldorado, Gloria, Isabel, Romeo, and yellow Stone while popular cultivars of *C. coronarium* L. include Nivea, Oreon, Albo, Golden Crown, Luteum, Tom Thumb Golden Gem, Tom Thumb-Primrose Gem and Tetra Comet. In addition to this, there is another type of annual Chrysanthemum, *i.e.* C. carinatum. Popular varieties of *C. carinatum* S. are Merry, Atrococcineum, Burridgeanum, Eclipse, Flammenspiel, John Bright and Northern Star.

## Propagation

Annual Chrysanthemums can be propagated by seeds. Their sowing time depends upon climate. It varies in different parts of India. In the Northern plains, winter season annual Chrysanthemums are sown during September–October. The seeds are sown in nursery beds, earthen pots, seed pans or wooden seed trays. The seed compost should consist of one part each of garden soil, coarse sand, farmyard manure and leaf-mould. For preparing the nursery beds, the soil should be dug up thoroughly and sufficient farmyard manure, should be mixed in soil. Raised nursery beds of convenient size (normally 60cm wide and 15cm high) should be prepared. If soil is heavy, some quantity of sand may be added. It is better if the soil of nursery bed or earthen pots is sterilized with 2% formalin. For this, soil is drenched with formalin solution and is covered with polythene sheet for 45hr. Afterwards, the polythene is removed and soil is dried before sowing the seeds.

Before sowing, the seeds should be treated with Cerasan (Methylmercury Chloride) (0.2%) and Captaf (0.2%) to prevent the seedlings from damping off disease. The seeds should be sown thinly and evenly as thick sowing causes damping off of seedlings. In nursery beds, the seeds are sown in rows spaced 6cm apart. Then, they are covered with finely sieved leaf-mould. Watering is to be done with a watering can having a fine rose both in beds and pots. In beds, when germination is over, water is given for proper moisture. Thereafter, the beds should be kept weed-free.

## Planting

The seedlings are transplanted 25 days after sowing at 4-leaf stage. Before transplanting, seedlings are hardened off by withholding water for 1 or 2 days or by exposing them gradually to sunlight. Transplanting is, generally, done either on a cloudy day or in the evening. Transplanting in evening is good as the night cool temperature is beneficial for the establishment. Light watering every day in early morning or late in the afternoon is required for about a week for proper establishment of the seedlings. Since annual Chrysanthemum is a tall plant, it should be planted 60cm apart.

## Manuring and fertilization

The farmyard manure or compost @ 3kg/$m^2$ is mixed in the soil. Chemical fertilizers—20g urea, 60–120g superphosphate and 30–60g murate of potash/ $m^2$ should also be added. Half quantity of urea and full of superphosphate and murate of potash should be applied at the time of bed preparation. The remaining quantity of urea must be applied one month after transplantation. Spraying plants with 2% urea twice or thrice is beneficial for good growth and flowering. Fertilizers should never come in the direct contact of the foliage since they cause scorching. Fertilizers should never be applied in the pot-grown annual flowers. However, some readymade pot-mixtures can be used. The pot-mixture should consist of 2 parts of garden soil and one part each of coarse sand and farmyard manure. Instead of fertilizers, it is better if pot-grown plants are given liquid feeding. The liquid manure is prepared by fermenting 1–2kg each of fresh cowdung and oil cake in 10 litres of water in a drum for one week. It is diluted to tea colour and sieved with the help of a muslin cloth. It is applied @ 500–1000ml/ pot at 7–10 days intervals.

## Aftercare

After transplanting, beds are weeded, hoed and watered regularly. As soon as seedlings are established in beds, pinching is done for making the plants bushy.

## Irrigation

Little water is needed everyday up to 7–10 days after transplantation. When the seedlings start new growth, profuse watering once or twice a week is required in beds. Later, frequency and quantity of watering depend upon soil and season. In lighter soils, more frequent irrigation is needed than that in heavy soils. The season of planting also determines the frequency of irrigation. During summer season, irrigation should be done at weekly intervals in beds, while at 10–12 days intervals in winters. Irrigation during rainy season depends upon prevailing weather conditions. Potted plants need daily watering during summer, whereas on alternate days in winter.

## Harvesting and Postharvest management

Most of the annual flowers are grown for garden display purpose in various ways. However, annual Chrysanthemums are grown commercially for cut flower or loose flower purpose. Their flowers are harvested when they are fully open and are sold in the local markets. The flowers, in general, are cut either late in the afternoon or very early in the morning. After harvesting, cut flowers should be put in a bucket of water filled up to one-fourth of the volume as it helps in their recovery from the shock of being cut away from the plant. As far as possible, the freshly opened flowers should be cut as freshness enhances their shelf-life. Since annual Chrysanthemums are used as cut flowers, proper postharvest management is necessary for prolonging their vase-life. The flowers are graded according to stem length, flower size, flower shape, flower color and freshness. The cut flowers/loose flowers are marketed in local markets. If flowers are not sold the day they are harvested, to store them in a cold storage is imperative.

## Seed standards

The percentage of minimum physical purity of foundation and certified seeds should be 98% with a minimum of 50% of germination capacity and 8% of moisture content. The presence of inert matter should not exceed 2.0%.

# 27

# Petunia (*Petunia* Juss. Solanaceae)

**Pre sowing seed treatment:** Fortification with GA3 100 ppm (or) moringa leaf extract 2% for 16 h (or)$KH_2PO4$ 2% for 16 hours

**Age of the seedling for transplanting:** The seeds are first sown in primary nursery (pot) and after 16 day the seedlings are transplanted to raised nursery and after 24 days transplanted to main field.

**Spacing :** 75 x 60 cm

**Fertilizer :** 125 : 75 : 150 kg NPK / ha

**Foliar spray :** $MgSO_4$ @ 2% spraying at 50% flowering stage

**Physiological maturity :** Seeds attain physiological maturity at 25 days after anthesis and it is associated with browning of pod and seed

**Harvest :** Pods are harvested in pickings on alternate days and upto 20 pickings pods can be used for seed extraction. Seed pods will ripen on the plant generally 4 weeks after pollination under day time temperature of 20 $^{o}$C to 29 $^{o}$C. The ripe seed pods are harvested by hand and allowed to air dry further in a warm, dry area.

**Grading:** Seeds are density graded with acetone and sinkers are selected for sowing.

**Storage:** Seeds are treated with Halogen mixture @3 g/kg or Diflubenzuron @ 1 ppm / kg and packed in aluminium foil poly laminated pouch.

# Section 3: Medicinal Plants

# 28

# Coleus forskohlii (Willd.) Bria. Libatae

**Varieties:** No named variety

**Soil:** Red sandy soil and sandy loam soil are highly suitable. Soil drainage is essential and hard pan and water logged soils should be avoided.

**Climate:** Suitable for plain and lower hills. Requires 70 cm annual rain fall

**Propagation:** Terminal three to four nodal cuttings measuring 10 cm length is used

**Land Preparation:** Apply 15 t FYM/ha. Ridges are formed at a spacing of 60 cm. Cuttings have to be planted at a spacing of 45 cm.

**Nutrient management:** Nutrients Quantity (kg / ha) N 30 P 60 K 50 The above nutrients can be applied in two split doses *viz.*, 30 and 45 days after planting. In micronutrient deficient soils, ZnSO4 can be applied as basal fertilizer.

**Irrigation:** Irrigate immediately after planting and later at weekly intervals. With hold the irrigation ten days before harvest.

**Plant protection:** The occurrence of wilt is becoming a major problem in coleus cultivation. Pathogens associated with wilt and root rot are *Fusarium chlamydosporum*, *Macrophomina phaseolina*, *Rhizoctonia solani* and Sclerotium sp. coupled with incidence of root knot nematode *Meloidogyne incognita*. Due to these diseases, the yield loss is upto 50 to 60%

## Management strategies for diseases

Select the coleus cuttings from disease free plants. Coleus cuttings have to be treated with Carbendazim solution (0.1%) before planting Soil drenching with Carbendazim (0.1%) or Propiconazole (0.1%) Soil application of FYM @ 12.5 ton/ha + 500 kg neem cake/ha + *Trichoderma viride* @ 2.5 kg /ha before planting is effective for biomanagment of nematode fungal disease complex involving *Meloidogyne incognita* and *Macrophomina phaseolina* Application of chemical nematicide Carbofuran 3G @ 1 kg a.i/ha before planting for control root-knot nematode. Use drip irrigation to minimize the spread of pathogens from infected plants to healthy plants

**Nematodes:** Dipping of stem cuttings in 0.1% *Pseudomonas fluorescens* at planting. Growing marigold (*Tagetus erecta*) as intercrop in between the rows of medicinal coleus and incorporateduring earthing up at 60-70 days after planting (or) Soil application of *Trichoderma viride* @ 2.5 kg/ha.

**Harvest:** Crop can be harvested six months after planting. Before harvest, top portion should be removed when sufficient moisture is in the soil. Roots are dug manually or by tractor drawn harvester. The soil particles are removed and the tubers are cut into small bits using motorized chopper to facilitate drying. The cut root bits are dried under sun for 3-5 days with frequent turnings until the moisture drops to 6-8 per cent.

**Yield:** Green roots: 15–20t /ha Dry roots: 2–2.5 ton/ha

# 29

# Ashwagandha (*Withania somnifera* Linnaeus) Dunal Solanaceae

## Botany

Ashwagandha (*Withania somnifera*), also known as Indian ginseng belonging to the family Solanaceae, is an important ancient plant, the roots of which have been employed in Indian traditional systems of medicine, Ayurveda and Unani to cure **asthma, fever, chest complaints.** It is an erect branching undershrub reaching about 1.50 m in height. It grows in dry and sub-tropical regions. Being hardy and drought tolerant species with its enormous biocompounds, its usage is forever regarded and continuous to enjoy the monopoly in many parts of India, particularly in Madhya Pradesh. It grows in dry parts in sub-tropical regions. Rajasthan, Punjab, Haryana, Uttar Pradesh, Gujarat, Maharashtra and Madhya Pradesh are the major Ashwagandha producing states of the country. Roots are fleshy, tapering, whitish brown. Leaves are ovate and flowers are greenish. The mature fruits are orange-red berries.

**Varieties**:Poshita and Rakshita are high yielding varieties released by CSIR-CIMAP, Lucknow. Jawahar 20 is cultivated in Madhya Pradesh. WSR is another variety released by CSIR-Regional Research Laboratory, Jammu. Nagori is a local variety with starchy roots

## Soil

Ashwagandha grows well in sandy loam or light red soil having pH 7.5 to 8.0 with good drainage. Black soils or such heavy soils are suitable for cultivation.

## Climate

It is grown as late rainy season (*kharif*) crop. The semi-tropical areas receiving 500 to 750 mm rainfall are suitable for its cultivation as rainfed crop. If one or two winter rains are received, the root development improves. The crop requires relatively dry season during its growing period. It can tolerate a temperature range of 20°C to 38°C and even low temperature as low as 10°C. The plant grows from sea level to an altitude of 1500 meter above sea level.

**Season:**It is planted late in the rainy season around August-September and harvested in the next May.

**Seed rate**: About 10 – 12 kg/ha of seed is required for broadcasting. For transplanting, seed rate of 5 kg/ha is required. The seedlings of six weeks old are transplanted at a spacing of 60X 60 cm.

**Germination improvement**:Mechanical scarification of seeds with sand for six minutes followed by soaking in $GA_3$ 500 ppm solution for five hours significantly improved the germination of seeds.

**Seed treatment :**Seed should be treated with Thirum or Dithane M45 at the rate of 3 g/kg seed before sowing to protect the seedlings from the seed borne diseases.

### Nursery raising

It is propagated by seeds. Fresh seeds are sown in well prepared nursery beds. Although it can be sown by broadcast method in the main field, transplanting method is preferred for better quality and export purpose. For export, a well maintained nursery is a prerequisite. The nursery bed usually raised from ground level is prepared by thorough mixing with compost and sand. About 5 kg of seeds are required for planting in 1 hectare of the mainfield. Nursery is raised in the month of June-July. Seeds are treated in carbendazim to control wilt and seed borne diseases. Seeds are sown just before the onset of monsoon and covered thinly using sand. The seeds germinate in 5 to 7 days. About 35 days old seedlings are transplanted in the main field.

**Spacing:** The seeds in the nursery beds are sown in lines spaced at 5 cm and covered with light soil

**Transplanting:** When the seedlings are 6 weeks old and sufficiently tall they are transplanted in 60 X 60 cm

**Manures and Fertilizers:** 12.5 t FYM and apply 12.5 Kg N: 25 Kg $P_2O_5$ /Ha

**Harvest:** Seeds attain physiological maturity at 35 days after anthesis when the colour of seeds changes from orange to deep red. Harvestable maturity of seed attains at 42 days after anthesis.

**Yield :**50 to 75 kg of seeds

# 30

# Periwinkle (*Catharanthus roseus* Linnaeus) G. Don Apocynaceae

**Botany:** Periwinkles (*Catharanthus roseus*) belonging to the family Apocynaceae are evergreen perennials that produce bright blue flowers. They grow best in well-draining, moist soil and partial sun to full shade. They work well as sprawling groundcover or border plants and bloom from early spring to early fall. Periwinkles are often grown as transplants, but they can be started from seed about 10 weeks before the last frost with the proper care.

**Varieties:** Nirmal, Dhawal and Prabal

**Soil and climate:** It is suited to all types of soil and tropical climatic conditions. Well distributed annual rainfall of 100 cm or more is ideal for raising as a rainfed crop.

**Seeds and planting:** Propagated through seeds either by direct sowing or throughtransplanting.

**Seed rate:** 2.5 kg/ha : Direct sowing

0.5 kg/ha : Through seedlings

45 - 60 days old seedlings are transplanted at a spacing of 45 x 20 cm during

June - July or September - October.

**Manuring**

**Basal:** Apply FYM at 10 t/ha and N P and K at 50 kg each /ha.

**Top dressing:** Apply 50 kg N 60 days after planting.

**After cultivation:** The crop requires 2 weedings, the first 90 days after sowing/ transplanting andsecond 60 days after the first weeding.

**Harvest:** The crop becomes ready for harvest of roots after one year. Two leaf strippings canbe taken, the first one after 6 months and the second after 9

months of sowing. Aerial parts arecut and the soil is ploughed for harvest of roots. Fruits are collected without damage.

**Irrigated Rainfed**

Roots 1500 kg/ha 750 kg/ha

Stems 1500 kg/ha 1000 kg/ha

Leaves 3000 kg/ha 2000 kg/ha

**Seed Production Technology:** The physiological maturity of periwinkle seeds is attained at 40 days after of anthesis with maximum dry weight, germination and vigour with a change of podcolour from green to yellow with translucence. Seeds can be processed by using 4/64" round perforated metal sieves with maximum seed recovery.

# 31

# Senna (*Cassia angustifolia* Vahl. Leguminasae)

**Botany**: Senna, *Cassia angustifola* belonging to the family Leguminosae is a perennial 1-2 m height undershrub. The stem is erect, smooth, and pale green to light brown with long spreading ascending branches. Leaves are compound with four to eight pairs of leaflets. The full grown leaflets are bluish-green to palegreen in colour and emit a characteristic fetid smell when crushed. The flowers are small and yellow and axillary arranged racemes. The pods are broadly oblong about 5-8 cm long and 2-3 cm broad, green in beginning and change to greenish-brown to dark brown on maturity. Each pod has 5 to 7 ovate, compressed, smooth, dark-brown seeds.

**Varieties :** KKM - 1, Anand Late Selection, ALFT-2 and Sona

**Soil and climate:** In Tamil Nadu, it is grown in sandy or sandy loam or laterite soils. It is a hardy warm weather crop grown under rainfed and irrigated conditions. Senna is also cultivated successfully on black cotton soils. It has a great tolerance to high soil salinity, however, shedding of some lower leaves often occurs. The soil pH suited for cultivation is 7.0-8.5. But, the most suitable to sandy loam to loam soils which are more friable and well drained.

**Seed rate:** 15 - 20 kg/ha. The seeds are scarified with sand or can be soaked overnight in water and sown in beds at a spacing 45 x 30 cm during February – March or June – July. Season: February - March and June - July.

**Manuring Basal:** Apply FYM 10 - 15 t/ha and N, P and K at 25, 25 and 40 kg/ha. Top dressing: Apply 25 kg N in two splits at 40 and 80 days after sowing.

**After cultivation:** One or two weedings.

### Plant protection - Pests

Aphids: Spray Dimethoate 30 EC or Methyl demeton 25 EC at 1 ml/lit of water. Harvest: The first harvest of leaves and pods are done 2 months after sowing and subsequent harvests at 30 days interval. Leaves and pods are dried for 7 - 10 days.

**Yield** - Irrigated Dried leaves : 2 t/ha. Dried pod : 150 - 200 kg/ha. Rainfed Dried leaves : 1 t/ha. Dried pods : 75 - 100 kg/ha.

## Seed Production Technology

Seeds attain physiological maturity at 40 days after anthesis, associated with colour change of pods from green to brown. Seeds can be processed by using 8/64" round perforated metal sieves. Hard seededness can be effectively overcome by acid scarification with commercial Sulphuric acid @ 100 ml/kg of seed for 10 minutes.

# 32

# *Phyllanthus amarus Schumach & Thonn* - Euphorbiaceae (APG: Phyllanthaceae)

## Botany

*Phyllanthus* is an herb common to Central and Southern India. It can grow to 30–60 centimeters in height and blooms with many yellow flowers. All parts of the plant are employed therapeutically. It is a small erect annual herb 10-60 cm tall widespread in tropical areas. Leaves are small elliptical - oblong and flowers are whitish green & minute. It bears ascending herbaceous branches. The bark is smooth and light green. The fruits are tiny, smooth capsules containing seeds.Kizhanelli is a circum-tropical weed, thriving well under tropical conditions. In India, it grows abundantly up to 700m altitude in Punjab, Haryana, Uttar Pradesh, Bihar, West Bengal, Orissa, Madhya Pradesh, Maharashtra, Andhra Pradesh, Karnataka, Tamil Nadu and Kerala.

**Major constituents** – Phyllanthin (0.4-0.5%) and Hypophyllanthin

**Uses-** Hepatitis B and Jaundice

## Varieties

Navyakrit (CIMAP) - High herbage and active constituents.

**Soil and climate**: Well drained Sandy loamy/black soil pH of 6.5 to 7.5 is preferred. It grows well as tropical and sub tropical rainfed crop.Plant does not grow properly under shade. It grows under semi-temperate to tropical conditions and under high rainfall. However, the plant rarely survives under dry or very low temperature conditions. Water logging does not show any lethal effect on this crop.

**Sowing:**Plough the land twice/thrice and level the top soil. About 1 kg of seeds is required for raising seedlings to plant in one ha. The seeds germinate in about a week and are maintained upto 20 days. Soaking the seeds in fresh water for 20-30 minutes before sowing would help in increased germination. Also seed treatment with GA3 200 ppm for 6 hours increased germination percentage.

**Spacing and Transplantation**:3 to 4 week old plantlets can be transplantedat a spacing of 10x15 cm, accommodating 8 lakhs seedlings/ha

### Manuring

To encourage good vegetative growth, apply 10-20 tonnes of FYM, 50 kg N, 25 kg each of $P_2O_5$ and $K_2O$ are applied per hectare. Half the dose of N, entire dose of P and K is applied as basal and the remaining N is applied in two split doses, the first at 30th day and the second at 60th day.

### Plant protection

Normally in this herb the pest and disease management efforts required are not high.

### Harvest

If cultivated in June to July, the plant is ready for cultivation in September to October. September harvesting is ideal for high Phyllanthin. From the date of the planting, in 80 to 100 days the plants grow to the maximum. At this time the concentration of active constituents in the leaves as well as the quantity of leaves are at highest. The leaves are dried in sun.

### Yield

An average yield of 17.5 tonnes of fresh herbage or 1750 kg/ha of dry herb is obtained.

# Section 4: Seed Spices

# 33

# Coriander (*Coriandrum sativum* Linnaeus Umbelliferae)

**Botany:***Coriandrum sativum* belonging to the family Umbelliferae. It is an annual soft shrub that attains a height of 1 to 3 feet. Leaves are in basal cluster varies from 3 to 15 cm in length and are oblong to ovate in shape. Leaflets are 1 to 2 centimeter in length and are ovate to round. Flower is of white color which can vary from light purple to creamish in color. These appear in umbles. Fruit is roundish in shape bearing color that varies from yellow to brown. It is tripped and gets divided in to two then crushed with fingers. The plant flowers and fruits in late winters.

**Land Requirement:** Land to be used for seed production of coriander should be free from volunteer plants

**Varieties:** CO 1, CO 2, CO 3, CO (CR) 4, GAU 1, UD 1, UD 2, UD 20 and UD 21

**Soil:** Well drained black cotton soil and red loamy soil. For rainfed cultivation soil should be clay in nature and the pH should be 6 – 8.Coriander performs well at a temperature range of 20 – 25 °C.

**Climate:** Cool and comparatively dry, frost free climate

**Season:** June - July and October - November.

**Seed rate:** 10 - 12 kg/ha (irrigated crop)

20 – 25 kg/ha (rainfed crop)

Whole seed will not germinate and hence the seeds are split open into halves before sowing for more germination percentage.

**Seed Treatment:** Soak the seeds in water for 12 hours. Treat the seeds with *Azospirillum* @ 1.5 kg /ha for better crop establishment + *Trichoderma viride* @ 50 kg/ha to control wilt disease.

Presowing seed hardening treatment with Potassium dihydrogen phosphate @ 10 g/lit of water for 16 hours is to be done for rainfed crop.

### Field preparation and sowing:

Prepare the main field to a fine tilth. Add FYM 10 t/ha before last ploughing. Form beds and channels (for irrigated crop). Sow the split seeds at a spacing of 20 x 15 cm. Spray pre-emergence herbicide Fluchloralin 700 ml in 500 lit of water per ha . The seeds will germinate in about 8-15 days.

**Isolation:** FS:200 m and CS:100 m

### Manuring

**Basal:** Apply 10: 40: 20 kg of NPK/ ha for rainfed and irrigated crops.

**Top dressing:** Top dressing may be done at 10 kg N/ha 30 days after sowing for the irrigated crop only.

**Irrigation:** First irrigation immediately after sowing and the second on the third day. Subsequent irrigations at 7-10 days interval.

### After cultivation

Thinning is done 30 days after sowing by keeping 2 plants per hill. Weeding is done as and when necessary. Spray CCC @ 250 ppm one month after sowing for inducing drought tolerance in rainfed crops.

### Plant protection

### Pests

**Aphid:** Spray Methyl demeton 25 EC @ 2 ml/lit or Dimethoate 30 EC @ 2 ml/ lit.

### Diseases

**Powdery mildew:** Seed treatment with *Pseudomonas fluorescens* (Pf 1) @ 10 g /kg and foliar spray of Pf1 2 g/lit or Spray Wettable sulphur 1 kg/ha or Dinocap 250 ml/ha at the time of initial appearance of the disease and 2nd spray at 10 days interval. Neem seed kernel extracts 5 % spray thrice (1st spray immediately after the appearance of disease, 2nd and 3rd at 10 days interval).

**Wilt:** Seed treatment with *Pseudomonas fluorescens* @10g /kg followed by soil application of Pf1 @ 5 kg /ha

**Grain mould:** Spray Carbendazim 0.1 % (500 g/ha) 20 days after grain set.

**Foliar spray:** 0.5 % Fe $SO_4$ at 30 and 45 DAS

## Field standards

| Factor | Foundation seed | Certified seed |
|---|---|---|
| Isolation distance | 200 m | 100 m |
| Off-types | 0.10 | 0.50 |
| *Objectionable weed plants | - | - |

*Objectionable weeds *Lathyrus*

**Physiological maturity:** 45 days after anthesis. Colour change from green to light olive brown.

## Harvest

The plants are pulled just when the fruits are fully ripe but green and start drying. The plants are dried and thrashed with sticks, winnowed and cleaned. For leaf, pull out the plants when they are 30-40 days old.

**Yield :** 300-400 Kg/ha

## Seed standards

The percentage of minimum physical purity of foundation and certified seeds should be 97% with a minimum of 65% of germination capacity and 10% of moisture content. The presence of inert matter should not exceed 3.0%.

# 34

# Fenugreek (*Trigonella foenum graecum* Linnaeus Papilionaceae)

**Botany:**Fenugreek is a leguminous plant, belonging to the Papilionaceae family. Fenugreek is an annual herbaceous plant between 20 and 50cm high. (sometimes up to 100cm.). Its stem is erect, branched, grooved and with little pubescence. The leaves are compound, petiolate, oblong leaflets formed by three kills reminding those of alfalfa. White or yellow flowers , isolated or arranged in pairs. Fenugreek fruits are narrow and elongated pods. They contain 10 to 20 seeds, tiny yellow, quadrangular, with a groove in the central part dividing them into two unequal lobes.

**Land Requirements:** Land to be used for seed production of Kasuri methi and methi shall be free from volunteer plants.

**Varieties:** Co 1, Pusa Early Bunching, Lam selection 1, Rajendra Kranti, Kissar Sonali, RMT1and CO 2.

**Soil:** A rich well drained loamy soil is best suited.

**Climate:** Cool and comparatively dry, frost free climate. It requires a warm climate with mild temperatures for a good seed production (between 8 and 27°C).It grows well in full sun. Sow it in places sheltered from the wind. Resistant to frost and temperatures down to -5°C. However, low temperatures during planting causes not all seeds to germinate.

**Season:** June - July and October - November.

**Seed rate:** 12 kg/ha.

**Seed treatment:** *Azospirillum* 1.5 kg + *Trichoderma viride* @ 50 g/ha for 12 kg of seeds.

**Field preparation and sowing:** Prepare the main field to a fine tilth.Add FYM 20 - 25 t/ha before last ploughing. Form beds and channels of 3.5 x 1.5 m.Seeds are sown at a spacing of 20 X 15 cm. Spray pre-emergence herbicide Fluchloralin 700 mlin 500 lit of water per ha.

**Isolation:** FS: 50m and CS: 25m

## Manuring

**Basal:** Apply 30:25:40 kg of N, P, K /ha.

**Top dressing:** Apply 20 kg of N at 30 days after sowing.

**Irrigation:** Give first irrigation immediately after sowing, second on the third day and subsequently at 7 - 10 days intervals.

**After cultivation:** Plants are thinned at 20 - 25 days after sowing and the thinned seedlings are used as greens.One pinching at a height of about 4" will encourage branching. Weeding is done as and when necessary.

## Plant protection

### Diseases

**Root rot:** Soil application of Neem cake @ 150 kg/ha and Seed treatment with *Trichoderma*

*viride*@ 4 g/kg or drenching with Carbendazim 0.5 g/l or Copper oxychloride 2 g/litre or

*Trichoderma viride* @ 5 kg/ha.

**Powdery mildew:** Dust Sulphur at 25 kg/ha or foliar spray with wettable sulphur 2 g/lit at thetime of appearance of disease.

## Field standards

| Factor | Foundation seed | Certified seed |
|---|---|---|
| Isolation distance | 50 m | 25 m |
| Off-types | 0.10 | 0.20 |
| *Objectionable weed plants | 0.010 | 0.020 |

**Harvest:**The crop is harvested when the plant is in bloom.The fruit is harvested when the pods have reached maturity, and many of them have acquired a curved shape resembling a sickle. Drying of seeds is done in full sun and in a well ventilated area.

**Yield:** 500 - 700 kg/ha.

## Seed standards

The percentage of minimum physical purity of foundation and certified seeds should be 98% with a minimum of 70% of germination capacity and 8% of moisture content. The presence of inert matter should not exceed 2.0%.

# Section 5: Plantation Crops

# 35

# Cocoa (*Theobroma cacao* Linnaeus Steruliaceae)

**Botany:** The Cocoa (*Theobroma cocoa*) belonging to the family Sterculiaceae. Cocoa tree is a shade tolerant, moisture loving, understory rainforest tree. The trees live for up to 100 years, but cultivated trees are considered economically productive for only about 60 years. Naturally Cocao grows to a height of 15 meters, but cultivated trees are trimmed shorter to make harvesting easier. The leaves of Cocoa are smooth bright green, oblong, about 15cm by 8cm. The seeds are encased in a large colourful pod which grows close to tree after a flower. The large pod is green while maturing and turns yellow, orange, red or purple when ripe. They range from about 10 cm to greater than 40 cm in length.

**Soil and climate:** Potash rich alluvial soils friable in nature with high humus and moisture retentivity with a pH of 6.6-7.0 are suitable. Cocoa is normally cultivated at altitudes upto 1200 m above MSL with an annual rainfall of 150 cm and a relative humidity of 80 % and annual mean temperature of 24°C to 25°C. Cocoa can be grown as intercrop in coconut and arecanut gardens.

**Varieties:** Criollo, Forestero and Trinitario. CCRP – 1, CCRP – 2, CCRP – 3, CCRP – 4, CCRP – 5, CCRP – 6 and CCRP – 7.

**Season:** June - July and September - October.

**Seeds and sowing:** Propagated by seeds. Before sowing the seeds the pulp adhering to the seeds has to be removed. Cocoa seeds are individually sown in polybags soon after extraction.

**Irrigation:** Irrigation should be given as and when necessary. During summer months irrigation should be given once in three days.

**Manuring:** Trees of 3 years of age and above are manured with 100 g N, 40g P and 140g K per tree in two split doses during April - May and August - September. Trees younger than three years may be applied with half of this dose.

**Formation pruning:** Done in young plants of cocoa (1 year after planting). The height of first jorquette is kept at 1-1.5m from the ground. Structural pruning: done generally 16-24 months after planting. Done to maintain tree at optimum height.

**Maintenance Pruning:** Starts from second year of planting. Remove low and hanging branches. Remove excess number of chupons regularly. Remove unproductive branches, dead, diseased and badly damaged branches in periodical intervals.

**After cultivation:** Weeding is done as and when necessary. The unproductive shoots, dead, diseased twigs should be removed periodically. Banana is better raised as a primary shade plant in the early years of plantation.

**Harvesting:**Bearing starts from 2nd year onward. Pods mature in about 5 to 6 months. Two crops, are there, first from Oct- Jan and 2nd from April-June. The ripen pods are harvested without causing injury to cushions from which they are developed. This is necessary as flowers are produced again from these "Cushions". The pods are opened by heating on a hard surface or using a mallet. Knife is not used as it may cut the beans inside. The beans are scooped out from the pod. To get 1 kg of dry cocoa beans about 15 to 30 pods are required.

**Yield:** 500-1000 kg of dry beans/ha.

# 36

# Cashew (*Anacardium occidentale* Linnaeus Anacardiaceae)

**Botany:** Cashew belongs to the Family Aricardiaceae. A small evergreen tree native of tropical America from Mexico to Peru and Brazil but now cultivated largely in Malabar, Kerala, Karnataka, Tamil Nadu and Andhra Pradesh and to some extend in Maharashtra, Goa, Odisha and West Bengal

The commercial production is mainly confined to India, Mozambique, Tanzania, Kenya; Brazil, Philippines, Malaysia and Sri Lanka.Cashew is an important foreign earning crop of India, remarking second (29%) in the international trade of the nuts. It is a crop of marginal lands and can be grown under rainfed conditions.

**Soil and climate:** It grows up well in all soils. Red sandy loam is best suited. Plains as well as hill slopes upto 600 - 700 feet elevation are suitable for cultivation. Season: June - December. Propagation: Soft wood grafting, air layer and epicotyl grafting. Requirement of plants: 200 plants/ha. Preparation of field: Pits of 45 cm3 size are dug and filled up with a mixture of soil + 10 kg FYM + one kg neem cake and 100 g Methyl parathion 1.3 %.

**Varieties:** VRI 1, VRI 2, VRI 3, VRI 4 and VRI (CW) H1

**Spacing:** 7 m either way.

**High Density Planting:** Spacing of 5 x 4 m accommodating 500 plants per hectare is recommended prune the interlocking branches during the July-August to maintain the frame.

## Manuring

| | FYM or Compost (kg) | N (g/ per tree | P (g/ per tree) | K (g/ per tree) |
|---|---|---|---|---|
| I year | 10 | 70 | 40 | 60 |
| II year | 20 | 140 | 80 | 120 |
| III year | 20 | 210 | 120 | 180 |
| IV year | 30 | 280 | 160 | 240 |
| V year | 50 | 500 | 200 | 300 |

Fertilizer application may be done during October - November in the East Coast areas. Wherever possible the fertilizer can be applied in 2 equal split doses during June-July and October-November periods under eastcoast area, a fertilizer schedule of 1000:125:250 g NPK/tree is recommended tree.

**Irrigation:** Noramally grown as a rainfed crop. Irrigation once in a west from flinching to fruit maturity stage is good to increase the yield.

**Intercropping:** Plough the interspaces after the receipt of rain and raise either groundnut or black gram till the trees reach bearing age.

**Training and pruning:** Develop the trunk to a height of 1 m by removing low lying branches. The dried twigs and branches should be removed every year.

**Harvest:** The plant starts yielding 3rd year onwards. The peak picking months are March to May. Good nuts are grey green, smooth and well filled. After picking, the nuts are separated from the apple and dried in the sun for two to three days to bring down the moisture content to 10 to 12%. Properly dried nuts are packed in alkathene bags. This will keep for 6 months.

**Yield:** 3-4 kg/tree/year.

# 37

# Coffee (*Coffea arabica* Linnaeus Rubiaceae)

**Botany:** Coffee trees are evergreen shrubs of the rubiaceae family. Coffee of commerce consists of seeds of coffee which when roasted, ground and soaked in hot water yields a fragrant stimulating infusion used for preparing the beverage called Coffee. It is native of America and introduced in India in 1936 near Chickmanglur in Karnataka.

Coffee plants have bright green opposite leaves with smooth margins. Its flowers, white, grow in clusters in the axils of the leaves and are aromatic. From them, fruits are born.They are red drupes with the size of a cherry The outside of the fruit is fleshy and inside it there are two seeds or coffee beans, surrounded by a membranous layer of leathery texture, hence this layer is commonly known as the "parchment". Some species of trees only produce one seed per fruit are known as "pearl coffee". In double seed varieties, the two seeds that grow inside each fruit press each other, so they stop the growth in the part which is in contact, so this becomes flat.

**Soil and Climate:** *Coffee arbica* comes up well in high altitude from 800 to 1650 meters but -C. *canephora Pierre ex A.Froehner* adopts well to lower elevations. At high altitude the crop is often late and susceptiable to frost and high winds.Evenly distributed rainfall of 2250 mm is essential. Heavy rainfall is not counducive as it encourages rotting of leaves and plants are easily susceptible to fungal diseases.

The temperature range is 50 to 80°F. It can also be grown at higher provided shade is there. In India, coffee is grown in hilly areas. There is a distinct dry period of about 90-100 days or from the Dec. March-April which is best for harvesting and processing and also for maturity of flower buds.The shallow soils with little organic matter to virgin soils on steep mountain slopes or almost flat lands are suitable for coffee. In Karnataka where *Coffee arabica* is grown on commercial scale the soils are red loamy and deep. Soils of "coffee estates are generally richJn' aluminum and iron content. The nitrogen percentage is also fairly high and low in lime per cent. The soilsare acidic with pH of 6.0.

**Propagation:** Propagated both by seeds and cuttings Seeds are collected during December. Seed should be heavy broad and boat shaped. Pulp of the seed is removed by hand: The fruits are rubbed with ash to prevent sticking of seeds and dried in shade and sown at 2.5 cm apart 4000 seeds are required5 for one ha Seeds germinate within 4 to 5- weeks. Then the seedlings are uprooted and transplanted in bags or in nursery bed at 25 cm apart.

**Vegetative Propagation:** Cuttings have to be selected from bushes of outstanding performance and made to root in the propagation chambers or in specially laid out nursery .beds. The cuttings root within.4-5 months. Rooting can be hastened, by employing growth regulators like IBA or by dipping the cuttings in cows urine.

**Planting:** Pits of 45 to 60 cm cube are dug during Jan.-Feb. at a spacing of 2.5x2.5m or 3x3 m depending on species to be planted. For *C. arabica* spacing of 2.5x2.5m is recommended and for *C. Canephora* spacing of 3x3m is adopted. Planting is done during June-July. In the initial stage staking is done for giving support to the; seedlings.

**Provision of Shade:** Prefers a subdued light *i.e.* partial shade where, sunlight is intense. The general practice, in East Africa-is to provide shade to the plantations grown below 1600 in- and: to grow coffee without shade above this altitude. Shade is necessary to maintain soil moisture and soil temperature. In Brazil coffee is grown without any shade. In India the environmental conditions are entirely different. The severe shade-often scorches the leaves and dries up the soil resulting in the loss of soil moisture and increase/ in soil temperatures which, adversely affect the coffee bush.

## Advantages of Shade Trees

1. They protect the bushes from torrential rains.
2. Provides a heavy mulch for the bushes.
3. Prevent soil erosion.
4. Adds organic matter to enrich fertility status of the soil.
5. Increase nitrogen content of the soil.

The following are some of the shade trees commonly used in coffee plantations, species of *Ficus, Terminale* sp. *Arfocarpus heteophytius, Greviied robustai* (Silver oak), *Albizy, Lebback, Erythina tillthesperna; Gliricidia maculate* and *Deucaene glauea.* Though the shade trees shed leaves and add humus to the soil, they compete for the nutrients with coffeebushes.

**Manuring:**The recommendations are based on the results of experiments, the nutrient removal and the productivity of coffee bushes under the different conditions of soil,climate and culture. The following fertilizer recommendations are given as the minimum requirements.

Fertilizer requirement of young plants from 1st to IIIrd year.

| Fertilizers | Kg/ha |
|---|---|
| Stermeal | 336 |
| Urea | 17 |
| Muriate of potash | 19 |
| Single super phosphate | 22 |

These fertilizers are applied in four split doses each of 106 kg in Feb./March, May/June, Aug./Sept, and Nov./Dec. The following fertilizer doses are recommended for mature bused of over three years age NPK 45:34:45 Kg/ ha.Foliar sprays of nutrients mixtures containing nitrogen, Phospheric acid, Potash, Magnesium and Zinc may be applied along with regular spray" programme.Foliar sprays of nutrients during the break in monsoon in July-Aug. have also been found to arrest fruit drop in coffee considerably. In problem areas where roots are unable to absorb sufficient nutrients from soil due to very low soil temp, lack of moisture of a resitricted injured or diseased root system, foliar feeding is very effective.

**Training and Priming:***Arabica* coffee is grown in India on single stem. The verticle growth of the plant is checked at two stages with the object of having good lateral spread and to secure a semispherical bearing surface. This object is achieved by topping and contering. The young tree is allowed to grow until it develops afcrown wood on the main stem and on the primary branches. When the stem mature, it is topped at a height of one meter. Timely topping of the young plant helps to develop a good framework on the bush. Centering is done in the 2nd or 3rd year to strengthen the stem and the primaries. Centering provides sufficient aeration to the lower region of the plant. Criss - cross and overriding shoots and unproductive wood are periodically removed, unproductive wood between all primaries and secondaries are also to be removed and only healthy vigorous growth is alone encouraged. These operations are generally carried in June - July and Sept. Oct.

**Reuvenation Prurning:**This consists of part of pole pruning generally done once in 4 or 5 years to bring back old coffee to shape again & control unwanted shoots. This is needed for the production of healthy vigonous cropping wood after heavy bearing or heavy defolation due to disease or debility. This type of pruning include removal of dead, exhausted, dried and worn out branches. This operation is taken up- immediately after harvest so that the plants will have the benefit of summer showers after prunning for the production of new flush.

**After Culture:**Digging and forking open are usually done during Sept. Nov. and again in Feb - March. This may be done annually but it is a costly item and also detrimental to feeding roots of the bush. Therefore, some plantor do digging once in 4 to 5 years weed control is done in intial years. Chemical weedicides like Dalapon (for controlling grasses and amino salt of 2, 4 - D) for controlling broad leaved weed are used.

Mulching and cultivation of green manuring plants are done with the object of adding dry matter to soil. Mulching also controls and check the weed growth.

**Harvest:** Harvest starts during October and extends upto February. Coffee fruits should be harvested as and when they become ripe. Coffee is just ripe when on gently squeezing the fruits the beans inside come out easily.

**Fly picking:**Small scale picking of ripe berries during October to February

**Main picking:**Well formed and ripened berries are harvested during December. Bulk of the yields are obtained from this picking.

**Stripping:** picking of all the berries left irrespective of ripening.

**Cleanings:** This is collection of fruits that have been dropped during harvesting. Unripe fruits should be scrupulously sorted out before using the fruits for pulping. They may be dried separately as cherry.

**Average Yield:** Arabica 400 to 650 kg/ha, Robusta 350 to 725 kg/ha.

# 38

# Coconut (Cocos nucifera Linnaeus Palmae)

**Botany:** Coconut palm (*Cocos nucifera*) belonging to the family Palmae is a slender palm tree widely distributed throughout the tropical regions,with a size that can vary between 12 to 30 meters high. Each inflorescence is a branched flower cluster (panicle), located at the base of the leaves. Its flowers, male and female (monoecious), can be white to pale yellow. Fruiting occurs throughout the year, and the fruit is a drupe with a dry fibrous mesocarp popularly known as coconut.

Coconut fruit has an ovoid shape, constituted by 3 angles and 3 germination holes at the end, with a size between 20 and 30 centimeters in diameter. Coconut fruit consists of a brown fibrous bark and a hard shell, booking a seed inside. The central cavity of the fruit has an aqueous or milky liquid called coconut water that is a nutritional mixture aimed for coconut to grow. When the fruit ripens , this fluid accumulates fat and it is called coconut juice or milk. The seed of the coconut palm is the second largest known , after the coco-de-mer (*Lodoicea maldivica* ), that can reach 40–50 cm in diameter and weigh from 15 to 30 kg. The seed of coconut is hollow inside and it is covered with a white pulp, oily and edible.

**Medicinal uses:**It is said to have analgesic, antirheumatic, antibacterial , anticancer, antioxidant, hypoglycemic, hypotensive and immune stimulating properties. Its oil also has numerous cosmetic applications

**Varieties**

Hybrids: VHC1, VHC2 and VHC3

Tall: VPM3, ALR 1, ALR 2 and West Coast Tall

Dwarf (tender coconut): COD, CYD, CGD and MYD

**Soil and Climate :**Coconut palm thrives in almost all types of wall drained soils such as coastal sand, red loam, laterals alluvial and reclaimed soils of marshy low lands. Coconut is a crop bumid tropics. Though it is mainly grown in the coastal plains it *is* possible to grown even at elevation of 600 to 900 m above M.S.L. in areas near the equator where the temperature remains favourable. Among the climatic. factors affecting the palm, rainfall is the most important;. A rainfall of 1000 to 2250 mm per annum evenly distributed throughout the year appears to be most congenial. Regions with long and pronounced dry spells are not suited to its growth. Coconut palm requires equable climate neither very hot not very cold. The maximum mean annual temp, for good growth is about 27°C with a diurnal vanetion of about 6° to 7°C. Persistant high humidity is harmful and incidence of bud rot is more under such a conditions. The palm require bright, sunshine of about 2000 hours a year.

**Planting Material (Propagation)** :Since, it is a cross-pollinated crop which is propagated only by seeds, the selection of planting material is of vital, importance selection has to be made at the mother palm level and at the seedling stage. The mother palm should be between age group of 25 to 60 years, should be healthy, high yielding and regular in bearing. Immature arid under-developed seed nuts should not be used. The selection of seedlings at nursery stage is also important. Generally nuts harvested from January to April are used for raising seedlings.

The seedlings should be.

1) Healthy
2) Should have minimum of 5 to 6 leaves when they are one year old.
3) The leaves should have been splited.
4) The girth of seedling at collar region should be more
5) Should have 5 to 6 roots.

**Planting seasons:** Jun - Jul and Dec - Jan. The planting can also be taken up in other seasons wherever irrigation and drainage facilities are available.

**Preparation of Land and Transplanting :** The depth of pit depends on soil type in sand loam soil pits of 1 x 1 x m is generally recommended. In laterite soils, the pits of 1.2%1.2 x 1.2 M are necessary. The pits are taken at the distance of 7.5 to 9 M apart thus accommodating 177 to 124 palm/ha. The-planting is done by square system, deep planting method is adopted. It is good practice to spread two layers of coconut husk at the bottom of the pits in areas where drought conditions prevail The seedling is placed at the center of the pit in such a way that the top of the husk is just visible from outside. The earth is well pressed down in order to keep the seedling firmly in position.

In well drained soil where water stagnation is not a problem transplanting is done at the beginning of the monsoon. In low lying areas plantingjs .done after monsoon. The trans planted, seedlings should be shaded and irrigated properly during summer. Irrigation with 45 litres of water once in four days has been found to be the optimum especially in sandy soils.

**Manure Fertilization and Intercultivation:** Application of fertilizers in general reduces the prehearing age of palms. The plant generally start bearing at the age of five or Seven years after planting and the stabilized yield is obtained from 10th year onward till the age of 60 years.Regular intercultivation and manuring is essential - for stepping up. and maintaining the productivity of palm.' Tillage including digging, ploughing the interspaces, making shallow basins with a radius of 2m and applying fertilizer. The C.F.CRX Kasargod has recommended an annual application of following nutrients/palm/y ear,

| Sr. No. | Type of plantation | N | P205 | K20 |
|---|---|---|---|---|
| 1. | Ordinary tall varieties | 0.5 kg | 0.32 kg | 1.2 kg |
| 2. | For hybrid and high yielding varieties | 1.0 kg | 0.50 kg | 2.0 kg |

Application of the annual dose of fertilizers in two or more splits had been found highly beneficial in increasing the yield and quality of nut. To obtain higher efficiency in the uptake of nutrients of fertilizers are to be applied in circular basis 20 to 25 cm deep and 1.5 to 1.8 m radius round the base of the palm.

**Inter and Multiple Cropping:** In pure coconut garden when palms are spaced at 7.5 m x 7.5 m, as much as 78% of the available area is not effectively utilized. It is also seen that a pure coconut grove utilizes only half, of the available light. Introduction of cacao, pineapple and pepper has been found to help increase the dry matter production. from 12.67% to 193% under Kerala conditions. In a crop combination involving coconut, diascorea, cacao and Pineapple 17,500 coconus, 100 kg of diascofes. tubers, 300 kg of dried cacao beans, and 2000 kg of pineapple, fruits were harvested from one ha. The increased income through multiple cropping is about 420% more than that obtained from a pure crop of coconut

**Production of Barren Nuts** :The phenomenon of the occurrence of barren nuts (without or with imperfectly developed Kernel) is very old. Only Certain trees in the coconut plantation produce large number of barren nuts. The nuts are generally oblong in shape and quantity of husk produce is very much less as compared to normal, nut. - The embryo in the barren nut is mostly absent or when present. It is in varying stage of decay. Fungal infection is also sometime noticed in the embryo resulting in the decay of the kernel and loss of water inside. In the barren nut cracking of shell is relatively more common.

1. Several causes for the phenomenon have been reported :
2. Due to defective fertilization resulting in malformation of embryo.
3. Nutritional deficiency in the palm.
4. Excessive bearing.

**Harvesting:** Coconut usually ripens in about 12 to 13 months after opening of the infloresecence. In order to get the maximum yield of copra and oil, only full matured nuts should be harvested (The loss of copra is 6%, 16% and 33% in 11 months, 10 months and 9 months old nuts, respectively). Similarly, the reduction in the oil percentage in 11, 10 and 9 months old nut being 5, 15 and 33% respectively. Generally, harvesting is done once in 45 to 60 days. Tender nuts which are in great demand as a delicious soft drink particularly in West Bengal and Maharashtra are best harvested at the age of six to seven months. A fully matured fruit will have a composition by weight of about 35% husk, 12% shell, 28% meat or Kernel and 25% water.

**Yield:**The average yield per ha varies from 10,000 to 14,000 nuts per annum. From a well maintained garden an annual yield of 25,000 nuts / ha per year can be obtained.

# Section 6: Minor Fruits

# 39

# Amla (Emblica Officinalis Gaertn. Euphorbiaceae)

**Botany:** *Emblica officinalis* belonging to the family Euphorbiaceae commonly known as Amla or Indian Gooseberry or Nelli is an important crop with high medicinal value. The fruits have the richest source of vitamin-C (700 mg per 100 g of fruits) and is considered to be good liver tonic. The various preparations using Amla include Chyavanprash, Triphala churna, Brahma Rasayana and Madumegha *churna*. The fruit is valued as an antiscorbutic, diuretic, laxative, antibiotic and anti-dysentric. Phyllemblin, obtainedfrom fruit pulp has been found to have mild depressent action on central nervous system. It has good demand from the industries for the preparation of various health care products also like hair oil, dye, shampoo, face creams and tooth powder.

**Soil:** Light as well as medium heavy soils except purely sandy soil is ideal for amla cultivation. The tree is well adopted to dry regions and can also be grown in moderate alkaline soils.

**Climate:** It is a tropical plant. Annual rainfall of 630-800 mm is ideal for its growth. The young plant upto the age of 3 years should be protected from hot wind during May-June and from frost during winter months. The mature plants can tolerate freezing temperature as well as a high temperature upto 46°C.

**Varieties:** The varieties recommended for cultivation are Banarasi, Chakaiya, Francis, NA-4 (Krishna) NA 5 (Kanchan), NA-6, NA-7, NA-10 and BSR-1 (Bhavanisagar).

**Cultivation**

**Pre sowing seed treatment:** Fresh seeds are stratified in sand moistened to 60% with $KNO_3$ @ 5 g/lit kept at 5°C for 10 days to remove the morphophysiological dormancy. Dry storage of fresh seeds for 10 months can also remove this dormancy.

**Propagation :**Amla is generally propagated by shield budding. Budding is done on one year old seedlings with buds collected from superior varieties yielding big sized fruits. Older trees or poor yielders can be changed into superior types by top working.

**Planting and Manuring :**The pits of 1m$^3$ are to be dug during May-June at a distance of 4.5 m x 4.5 m spacing and should be left for 15-20 days exposing to sunlight. Each pit should be filled with surface soil mixed with 15 kg farm yard manure and 0.5 kg of phosphorus before planting the budded seedling.200 Seedlings per acre to be planted. Apply 90: 120: 48 NPK Kg per acre. Application of 30 g of nitrogen each year during September - October upto 10 years for each tree is recommended.

**Irrigation :**Young plants require watering during summer months at 15 days interval till they are fully established. Watering of bearing plants is advised during summer months at bi-weekly interval. After the monsoon rains, during October - December about 25-30 litres of water per day per tree through drip irrigation should be given.

**Training and pruning :**Leaving only 4-5 well shaped branches with wide angle at about 0.75 m from the ground level, other dead, diseased, week crisscrossing branches and suckers should be pruned off at the end of December.

**Mulching and Intercropping :** During summer, the crop should be mulched with paddy straw or wheat straw at the base of the tree upto 15-20 cm from the trunk. Inter crops like green gram, black gram, cowpeaand horse gram can be grown upto 8 years.

### Plant protection

Major insect : Bark Eating Caterpillar (*Inderbella tetronis*)

Major disease : Rust (*Ravenellia emblicae*)

### Control measures

Injection of Endosulphon 0.05% or Monocrotophos 0.03% in holes and plugging with mud is effective in protecting the tree against bark eating caterpiller.Spraying of Dithane M-45 @ 0.3% twice first in early September and second 15 days after first application controls the spread of rust.

**Flowering season**: January - February

**Fruiting season**: November – February

**Harvesting :**Amla tree starts bearing after about 4-5 years of planting. The fruits are harvested during February when they become dull greenish yellow from light green. Seeds attain physiological maturity 22 to 24 weeks after breaking fruit dormancy of 5 months when the fruit colour turns to yellowish green and seed colour turns to chestnut brown.

**Seed extraction:** Amla seeds can be extracted by soaking the fruits in 30% brine solution during night followed by drying during the day and repeated for 3 days. The mesocarp following soaking in brine solution remained fresh and can be used for pickle/byproducts preparation.

**Grading:** Using 8/64" round perforated metal sieve with maximum seed recovery can be used to process seeds. The size-graded seeds can be further upgraded by density grading using water to remove the light weight empty seeds.

**Storage:** Seed treated with Carbendazim 50 % WP @ 2 g /kg seed and halogen formulation (bleaching powder + $CaCO_3$ + Arappu *Albezia amara* leaf powder @ 5:4:1) @ 3 g/kg seed can be stored upto 18 months in cloth bag and at 24 months in moisture proof container. Seeds can also be treated with Carbendazim @ 4 g + Carbaryl 400 mg / kg and packed in 700 gauge polythene bag and stored in 50C.

**Yield**:A mature tree of about 10 years will yield 50-70 kg of fruit. The average weight of the fruit is 60-70 g and 1 kg contains about 15-20 fruits. A well maintained tree yields upto an age of 70 years.

# 40

## Jamun (*Syzygium cumini* (Linnaeus) *Skeels Myrtaceae)*

**Introduction:**Jamun (*Syzygium cumini)* belonging to the family Myrtaceae is a popular indigenous fruits Of India. It is also known as black plum, Indian black cherry, Ram jamun etc. in different parts of India. The tree is tall and handsome, evergreen, generally grown for shade and windbreak on roads and avenues. In India, the maximum number of jamun trees are found scattered throughout the tropical and subtropical regions. It also occurs in the lower range of the Himalayas up to an elevation of 1,300 meters and in the Kumaon hills up to 1,600 meters. It is widely grown in the larger parts of India from the Indo-Gangetic plains in the North to Tamil Nadu in the South.

**Soil:**The jamun tree can be grown on a wide range of soils. However, for high yield potential and good plant growth, deep loam and a welldrainedsoil are needed. Jamun can grow well under salinity and waterlogged conditions too.

**Climate:** Jamun prefers to grow under tropical and subtropical climate. It is also found growing in lower ranges of the Himalayas up to an altitudc of 1300 meters. The jamun requires dry weather at the time off towering and fruit setting.

**Varieties:** The common variety grown under North Indian conditions is "Ram Jarnun". It produces big sized, oblong fruits, deep purple or bluish-black in color at full ripe stage. The pulp of the ripe fruit is purple pink and the fruit is juicy and sweet. The stone is small in size. The variety ripens in the month of June- July and it is very common both in rural as well as in urban markets.

**Propagation:**The jamun is propagated both by seed and vegetative methods. Due to existence of polyembryony, it comes true to parent through seed. Though vegetative methods followed in most cases have attained some success, seed propagation is still preferred. However, seed propagation is not advisable as it results in late bearing.The seeds have no dormancy. Fresh seeds can be sown. Germination takes place in about 10 to 15 days. Seedlings are ready for transplanting for the use as rootstock in the following spring (February to March) or monsoon *i.e.* August to September.

Propagation of jamun is economical and convenient. Budding is practiced on one year old seedling stocks, having 10 to 14 mm thickness. The best time for budding is July to August in low rainfall areas. In the areas where rains start easily and are heavy, budding operations are attempted early in May-June. Shield, patch and forkert methods of budding have proved very successful. The possibility of better success has been reported in forkert method compared to shield or 'T' budding.

Jamun can also be propagated by inarching but it is not adopted commercially. About 60% air layers are obtained with 500 ppm IBA in lanolin paste, provided air layering is done in spring and not in the rainy season.Better rooting through cutting is obtained in Jamun under intermittent mist. Semi-hardwood cuttings of both *S. jambos* and *S.javanica*, 20-25 cm long, taken from the spring flush and planted in July treated with 2000 ppm IBA (indole butyric acid) give better results.

**Planting :**Jamun is an evergreen tree and can be planted both in spring i.e. February -March and the monsoon season *i.e.* July-August. The latter season is considered better as the trees planted in February- March have to pass through a very hot and dry period in May and June soon after planting and generally suffer from mortalities from the unfavourable weather conditions.

Prior to planting, the field is properly cleared and ploughed. Pits of 1 x 1 x 1 m size are dug at the distance of 10m both ways. Usually, work of digging of pits is completed before the onset of monsoon. The pit are filled with mixture of75% top soil and 25% well rotten farmyard manure or compost.

**Fertilizer Application:**The jamun trees are generally not manured. An annual dose of about 19 kg farmyard manure during the pre-beating period and 75 kg per tree bearing trees is considered.Normally, seedling jamun trees start bearing at the age of 8 to 10 years while grafted or budded trees come into bearing in 6 to 7 years. On very rich soils, the trees have a tendency to put on more vegetative growth with the result that fruiting is delayed. When the trees show such a tendency, they should not be supplied with any manure and fertilizer and irrigation should be given sparingly and withheld in September-October and again in February-March. Flowering season –June, Fruiting season July - August

**Irrigation:**In early stages, the jamun tree requires frequent irrigations but after the trees gets established, the interval between irrigations can be greatly decreased. Young trees require 8 to 10 irrigations in a year. The mature trees require only about half the number, which should be applied during May and June when the fruit is ripening.

**Intercropping:**In the initial years of planting, when a lot of interspace is available in the orchard, appropriate intercrop especially legummous crops and vegetables can be taken during rainy season.

**Training and Pruning:**Regular pruning in jamun is not required. However, in later years the dry twigs and crossed branches are removed. While training the plants, the framework of branches is allowed to develop above 60 to 100 cm from the ground level.

**Insect Pests:**Among the pests, white fly and leaf eating caterpillar cause great damage to the tree.

1. **White fly** *(Dialeurodes eugenia)* **:** Affected fruits get wormyappearance on the surface. It can be controlled byMaintain sanitary conditions around the tree, pluck all affected fruits and destroy them and dig up the soil around the tree trunk so that the maggots in the affected fruits and pupae hibernating in the soil are destroyed.
2. **Leaf eating caterpillar** *(Carea subtillis)* **:**This caterpillar is only found in Coimbatore. The insect infests the leaves and may defoliate the tree. It can be controlled by spraying Rogor 30 EC or malathion @ 0.1 per cent.
3. **Other pests :**Besides the above insects, the jamun crop is seriously damaged by pests like squirrels and birds like parrots and crows. These have to be frightened away by beating the drums or flinging stones.

**Diseases :**Among the diseases, the fungal disease anthracnose is notable .

1. **Anthracnose** *(Glomerella cingulata)***:**The fungus incites leaf spots and fruit rot. Affected leaves show small scattered spots, light brown or reddish brown in colur. Affected fruits show small water soaked, circular and depressed lesions. Ultimately, the fruits rot and shrivel.Spraying with Dithanc Z- 78 @ 0.2% or Bordcaux mixturc at: 4:4:50 conccntration shall check the disease.

**Harvest:** Jamun seeds attain physiological maturity 11 weeks after anthesis when the fruit colour changes to purplish black. Seed extraction: The collected fruits are heaped for one day and squeezed. Seeds will come out easily.

**Grading:** Fresh seed can be size graded using 20/64" round perforated metal sieve with maximum seed recovery.

**Storage:** It is a recalcitrant seed and will lose viability upon storage due to desiccation (drying). When stored seeds are to be used for sowing, the dead seeds can be removed by density grading using water. The seeds lose viability completely within one month in tropical condition when the moisture content of seed falls below 20%. The critical moisture for safe storage is around 45%. Packing of seeds in polythene bags containing 2% moist sand stored at 10æ%C in refrigerator storage prolongs viability upto 3 months

**Yield :** The average yield of fruits from a full grown seedling tree is about 80 to 100 kg and from a grafted one 60 to 70 kg per year.

# References


Agrawal, R. L. 1980. *Seed Technology*. Oxford & IBH Publishing Co. PVT. Ltd. agritech.tnau.ac.in

Bhaskaran, M., K.Vanangamudi, A.Bharathi, P.Natesan, R.Jerlin, N.Natarajan and K.Prabhakaran. 2003. *Principles of Seed Production and Quality Control*. Kalyani Publishers, Ludhiana.

IFOAM Training Manual for Seed Saving, Compiled by the Centre for Indian Knowledge Systems, Chennai for International Federation of Organic Agriculture Movements (IFOAM) Bonn (Germany). pp. 123

Package of Organic *Practices from Tamil Nadu for Rice, Groundnut, Tomato and Okra*, 2006. Prepared by the Centre for Indian Knowledge Systems, Chennai for the National Centre for Organic Farming (NCOF), Government of India and Food and Agriculture Organisation (FAO), United Nations. pp. 174

Quality *Seed Production in Brinjal* (Tharamana Kathari Vidhai Vurpathi Muraigal). The Department of Seed Science and Technology, Tamil Nadu Agriculture University, Coimbatore. pp. 34

Quality *Seed Production in Chilli* (Tharamana Milagai Vidhai Vurpathi Muraigal). The Department of Seed Science and Technology, Tamil Nadu Agriculture University, Coimbatore. pp. 33

Quality *Seed Production in Ladies finger* (Tharamana Vendai Vidhai Vurpathi Muraigal). The Department of Seed Science and Technology, Tamil Nadu Agriculture University, Coimbatore. pp. 33

Quality *Seed Production in Pumpkin* (Tharamana Paranki Vidhai Vurpathi Muraigal). The Department of Seed Science and Technology, Tamil Nadu Agriculture University, Coimbatore. pp. 28

R. Abarna Thooyavathy, K. Perumal, V. Suresh and K. Vijayalakshmi. (2013).*Seed Production Techniques for Vegetables*. Centre for Indian Knowledge Systems (CIKS) Seed Node of the Revitalising Rainfed Agriculture Network.

Singh, P.M., Singh, B., Pandey, A.K. and Singh, R. (2010). *Vegetable Seed Production* - A Ready *Reckoner*. Technical Bulletin No.37, IIVR, Varanasi.

Sridhar, S. Arumugasamy, S. and Saraswathy, H. and Vijayalakshmi, K. 2003. *Organic Vegetable Gardening*, Centre for Indian Knowledge Systems, Chennai, pp.46

Training Manual on Principles of Seed Technology. The Department of Genetics and Plant Breeding, Tamil Nadu Agriculture University, Coimbatore. pp. 163.

Vanangamudi, K., Natarajan, N., Srimathi, P., et al. 2012. *Advances in Seed Science and Technology* (Vol 2) – *Quality Seed Production in Vegetables,* Agrobios (India), Jodhpur. pp.925

www.indiaagronet.com

www.indianspices.com

www.seedtamilnadu.com

www.tnau.ac.in

www.tngov.in

www.agriinfo.in
http://72.41.96.132/index.php/site/seeddepartment
agricoop.nic.in/.../seed/INDIAN_MINIMUM_SEED_CERTIFICATION_
agritech.tnau.ac.in/horticulture/horti_index.html